PHYSICS RESEARCH AND TECHNOLOGY

SELECTED ASPECTS OF THEORETICAL PHYSICS AND ORGANIC CHEMISTRY

PHYSICS RESEARCH AND TECHNOLOGY

Additional books in this series can be found on Nova's website under the Series tab.

Additional E-books in this series can be found on Nova's website under the E-books tab.

CHEMISTRY RESEARCH AND APPLICATIONS

Additional books in this series can be found on Nova's website under the Series tab.

Additional E-books in this series can be found on Nova's website under the E-books tab.

PHYSICS RESEARCH AND TECHNOLOGY

SELECTED ASPECTS OF THEORETICAL PHYSICS AND ORGANIC CHEMISTRY

V. BABKIN
G. ZAIKOV
V. TITOVA
T. PERESYPKINA
AND
O. PONOMAREV

Nova Science Publishers, Inc.
New York

Copyright © 2011 by Nova Science Publishers, Inc.

All rights reserved. No part of this book may be reproduced, stored in a retrieval system or transmitted in any form or by any means: electronic, electrostatic, magnetic, tape, mechanical photocopying, recording or otherwise without the written permission of the Publisher.

For permission to use material from this book please contact us:
Telephone 631-231-7269; Fax 631-231-8175
Web Site: http://www.novapublishers.com

NOTICE TO THE READER
The Publisher has taken reasonable care in the preparation of this book, but makes no expressed or implied warranty of any kind and assumes no responsibility for any errors or omissions. No liability is assumed for incidental or consequential damages in connection with or arising out of information contained in this book. The Publisher shall not be liable for any special, consequential, or exemplary damages resulting, in whole or in part, from the readers' use of, or reliance upon, this material. Any parts of this book based on government reports are so indicated and copyright is claimed for those parts to the extent applicable to compilations of such works.

Independent verification should be sought for any data, advice or recommendations contained in this book. In addition, no responsibility is assumed by the publisher for any injury and/or damage to persons or property arising from any methods, products, instructions, ideas or otherwise contained in this publication.

This publication is designed to provide accurate and authoritative information with regard to the subject matter covered herein. It is sold with the clear understanding that the Publisher is not engaged in rendering legal or any other professional services. If legal or any other expert assistance is required, the services of a competent person should be sought. FROM A DECLARATION OF PARTICIPANTS JOINTLY ADOPTED BY A COMMITTEE OF THE AMERICAN BAR ASSOCIATION AND A COMMITTEE OF PUBLISHERS.

Additional color graphics may be available in the e-book version of this book.

Library of Congress Cataloging-in-Publication Data

Selected aspects of theoretical physics and organic chemistry / editors, V. Babkin ... [et al.].
   p. cm.
 Includes index.
 ISBN 978-1-61728-684-1 (softcover)
 1. Chemistry--Statistical methods. 2. Statistical physics. 3. Correlation (Statistics) I. Babkin, V. A.
 QD39.3.S7S45 2010
 540.72'7--dc22
                     2010025949

Published by Nova Science Publishers, Inc. † New York

This book is dedicated to the memory of Frank Columbus

On December 1st 2010, Frank H. Columbus Jr. (President and Editor-in-Chief of Nova Science Publishers, New York) passed away suddenly at his home in New York.

We lost our colleague, our good friend, a nearly perfect person who helped scientists from all over the world. Particularly Frank did much for the popularization of Russian and Georgian scientific research, publishing a few thousand books based on the research of Soviet (Russian, Georgian, Ukrainian etc.) scientists.

Frank was born on February 26th 1941 in Pennsylvania. He joined the army upon graduation of high school and went on to complete his education at the University of Maryland and at George Washington University. In 1969, he became the Vice-President of Cambridge Scientific. In 1975, he was invited to work for Plenum Publishing where he was the Vice-President until 1985, when he founded Nova Science Publishers, Inc.

Frank Columbus did a lot for the prosperity of many Soviet (Russian, Georgian, Ukrainian, Armenian, Kazakh, Kyrgiz, etc.) scientists publishing books with achievements of their research. He did the same for scientists from East Europe – Poland, Hungary, Czeckoslovakia (today it is Czech Republic and Slovakia), Romania and Bulgaria.

He was a unique person who enjoyed studying throughout the course of his life, who felt at home in his country which he loved and was proud of, as well as in Russia and Georgia.

There is a famous Russian proverb: "The man is alive if people remember him." In this case, Frank is alive and will always be in our memories while we are living. He will be remembered for his talent, professionalism, brilliant ideas and above all – for his heart.

Dr. V. Babkin
403346, Volgograd province t. Mikhailoka
Street Democraticheskaya 85

Dr. G. Zaikov
117334, Moscow
Kosygin Street 4 Street

Dr. V Titova
Volgograd State Technical University
Volgograd Blvd Lenin 28
403811, Chair Organic Chemistry

Dr. T. Peresypkina
403343, Volgograd province t. Mikhailovka
Oborona st. 57/75

Dr. O. Ponomarev
Pensioner
109398, t. Moscow
Kuhmitirova 18/28

CONTENTS

Preface		ix
Chapter 1	The Method of Correlation Functions in the Problem of Many Objects	1
Chapter 2	The Method of the Constant Self-Consistent Field Used for the Decision of the Problem of Many Solids	9
Chapter 3	Dynamic and Stochastic Systems	15
Chapter 4	The Method of Approach MD	23
Chapter 5	The Method of Approach MD (Continuation)	29
Chapter 6	The Generalized Model of Matsubara	35
Chapter 7	Research of Nonequilibrium Dynamic Problems	41
Chapter 8	Conductivity in Polymers	47
Chapter 9	Conductivity in Polymers (Continuation)	53
Chapter 10	A Spectrum of Elementary Electronic Conditions in Polyarylenphtalydes with Two Soliton Excitations	61
Index		75

PREFACE

This book presents and discusses modern problems of organic chemistry and theoretical physics including the method of correlation functions of many objects; dynamic and stochastic systems; the generalized model of Matsubara; conductivity in polymers; research of nonequilibrium dynamic problems; and conductivity in polymers.

Chapter 1

THE METHOD OF CORRELATION FUNCTIONS IN THE PROBLEM OF MANY OBJECTS

1.1. INTRODUCTION

The Method of correlation functions (CF) or Functions of Grin (FG) brings together a linear system of equations. The Number of the equations is infinite. The Endless system of equations used is very difficult. The System is limited. All of CF (the order which above definite sign) are lowered completely or the senior CF is expressed using the younger CF. The System of linear equations is changed to a simpler system of equations. Both ways bring fair results. Such CF describes only their asymptotic forms on greater times.

Such ways of the breakaway of the endless chain are little used for systems with intermediate and tight connections (CF mainly contributes the area of small times). If memory about motion in accordance with dynamic equations is not quite lost, the contribution of the effects of dynamic relationships are taken into account. The times are different for different systems. The way to account for such effects is described below.

This way will approximate the behavior of CF on small times and moves over to the known method of the constant self-coordinated field in the event of a weak connection. The Physical sense of the new method consists in the variable self-coordinated field of entering. We reduce the endless system of equations to a final nonlinear system. Typical particularities appear during study of the nonlinear equations: the structure of phase space can turn out to be essential and be determined by image at calculation of the energy of elementary excitement, thermodynamic and kinetic features; "isle" dynamic exists in phase space in the field of stochastic behavior. This creates an "echo" type of effect. All this is greatly at tight binding.

1.2. THE METHOD CF IN THE PROBLEM OF MANY BODIES

Studying the problem of many bodies is a primary task of statistical mechanical engineers who must search for efficient methods for the calculation of equilibrium and nonequilibrium features of the system, consisting of a large number of particles.

We again mention the general methods of statistical physicists: CF and FG. Characteristics of systems of many particles (the energy spectrum, thermodynamic functions) are expressed through CF. CF is average importance on statistical ensemble from product with a certain number of operators. Average importance of any dynamic variable A are expressed as:

$$A = \frac{Sp A \rho}{Sp \rho} = <A>$$

where a matrix of a density (ρ) satisfies the equation

$$i \frac{\partial}{\partial \tau} \rho = [H, \rho],$$

$H-$ full Hamiltonian of system, $\hbar = 1$. This is a quantum equation of Liuvillya in operator form. In equilibrium position:

$$\rho = \rho_0 = \frac{\exp(-\tilde{H}\beta)}{Sp \exp(-\tilde{H}\beta)},$$

where $\hat{H} = H - \mu N$, $\beta = 1/kT$, $\mu -$ chemical potential.

Two-temporary CF are often used

$G(t,t') = <A(t)B(t')>$,

where A(t), B(t) - importance of operator A and B (by Heisenberg)

$$A(t) = \exp(i\widetilde{H}t) A \exp(-i\widetilde{H}t)$$

$G(t,t') = G(t-t')$ (external fields are absent) (1)

If A and B are an operator of the birth and deletion of the particle, poles (1) define the spectrum of elementary one-partial excitement. Type (1) is defined by the Hamiltonian type of system. Elektron-phonon systems with Hamiltonian will be explored in given work

$$H = \sum_{k\sigma} \frac{k^2}{2m} a^+_{k\sigma} a_{k\sigma} + \sum \omega_q b^+_q b_q + \sum A_q a^+_{k1\sigma} + a_{k2\sigma} r_q, \quad (2)$$

$r_q = b_q + b^+_{q'} a^+_{k\sigma'} a_{k\sigma'} b^+_q b_q$ - as operators of the birth and destruction of electrons and phonons accordingly (Frelih Hamiltonian). If dependency of the fraudulent integral from the nucleus coordinates is taken into account, (2) is converted:

$$H = \sum a^+_n H_n a_n + \sum a^+_n B_{nm} a_{m'}$$

$$B_{nm} = B_{nm}(...b_{q'} b^+_q ...) , \quad (3)$$

$$H_n = E_n + \sum \frac{1}{M!} F^{(n)}_{q1...qn} \prod_{i=1}^{M} r_{qi} + \sum \omega_q b^+_q b_q ,$$

where a_n^+, a_n - as operators of the birth on node n.

The importance of CF allows us to completely describe the thermodynamics of many-particle systems if times coincide. The System with Hamiltonian (2) shall consider. Averaging by statistical ensemble of Gibbs use

$$E = <H> = \sum \frac{k^2}{2m} <a^+_{k\sigma}(0)a_{k\sigma}(0)> + \sum(\sum \omega_q <b^+_q b_q> + \sum A_q <a^+_{k1\sigma}(0)a_{k2\sigma}(0)r(0)>).$$

It is necessary to know three functions of the distribution for calculation of the average energy

$$G_{kk\sigma}(0) = <a^+_{k\sigma}a_{k\sigma}>, \quad N_q(0) = <b^+_q b_q>,$$

$$M_{qk1k2\sigma}(0)G_{k1k2\sigma}(0) = <a^+_{k1\sigma}(0)a_{k2\sigma}(0)r_q(0)>$$

The Average energy allows the restoration of the rest of the thermodynamic functions. Knowledge of CF is necessary to calculate this.

The Method CF or FG is also used at other times. If System is found in a weak external floor (the detour from balances little), the reaction of the system on the external influence is possible to express using CF. The Advantage given way: do not use kinetic equation. The External field $V(t) = -AF(t)$ add to H:

$$i\frac{\partial}{\partial \tau}\rho = (H+V)^x \rho(t),$$

where the symbol $(.....)^x$ makes sense $a^x b = [ab - ba]$ (by Cube).

A - an operator, F(t) describes the dependency of the indignation from time. The Linear additive on external field use, $\rho(-\infty) = \rho_0$:

$$\rho(t) = \rho_0 - i\int_{-\infty}^{t}[\exp(-i(t-t')H^x)]V^x(t')\rho_0 dt', \qquad (4)$$

where $\exp \exp(a^x)b = \exp(a)b\exp(-a)$. From (4) get for average importance of the dynamic value B:

$$<B> = <B> - i\int_{-\infty}^{t}dt'F(t') <[B(t-t')'A]>.$$

For periodic external power $F(t) = F_0 cos\omega t$:

$$<\Delta B(t)> = <B> - <B>_0 = \text{Re}\{X_{AB}(\omega)F_0 e^{i\omega t}\},$$

where

$$X_{AB}(\omega) = \int_0^\infty \varphi_{AB}(t)e^{i\omega t}dt, \quad \varphi_{AB}(t) = i<[B(t), A(0)]>.$$

The Identity use:

$$[\rho_0, A] = i\int_0^\beta d\lambda \rho_0 \dot{A}(-i\lambda) \tag{5}$$

we shall convert $\varphi_{AB}(t)$ to type

$$\varphi_{AB}(t) = \int_0^\beta d\lambda <\dot{A}(i\lambda)B(t)>,$$

The formula of Cube get

$$X_{AB}(\omega) = \int_0^\infty e^{-i\omega t}dt \int_0^\beta <\dot{A}(-i\lambda)B(t)>d\lambda \tag{6}$$

For electrical conductivity tensor

$$\sigma_{AB}(\omega) = \int_0^\infty e^{-i\omega t}dt \int_0^\beta d\lambda <J_\nu(-i\lambda)J_\mu(t)>,$$

where J_V – components of the current (by Heisenberg) (v,µ = x, y, z). (n = 1) – for electro conductivity, (n = 2) – for thermal conductivity, (n = 3) – for differential thermal voltage, (n= 4) – for Peltier coefficient. Tensors of rank 2 are necessary to calculate:

$$L_{V\mu}^{(n)} = \int_0^\infty dt \int_0^\beta d\lambda < A_\mu^{(n)}(0) B_V^{(n)}(t+i\lambda) >,$$

$$A^{(1)}(t) = A^{(3)}(t) = B^{(1)}(t) = B^{(2)}(t) = J(t),$$

$$A^{(2)}(t) = A^{(4)}(t) = B^{(3)}(t) = B^{(4)}(t) = J_E(t),$$

J, J_E - operators of current and power flow (by Heisenberg).

If the external field is not small, the decision to write ρ is possible as exponent:

$$\rho(t) = \rho_0 - i \int_{-\infty}^t T \exp\left\{-i\int_1^t ds(H+V(s))\right\}^{|X} V^X(t')\rho_0 dt'.$$

we get

$$< \Delta B(t) = -i \int_{-\infty}^t F(t')dt' < [B(t,t'), A] > \qquad (7)$$

where

$$B(t,t') = T \exp\left\{i \int_{t'}^t [H+V(s)]ds\right\}^{|X} B.$$

(5) we use, (7) convert to type (8)

Dual-time CF is calculated in all events, with the exclusion of a nonlinear response. More Complex CF is required for a nonlinear response.

Two main methods for calculation of CF or FG exist: the method of theories of the indignations (the method of the diagrams) and the method of equation of the motion. Here we use the second method.

CF is required to calculate

$G(t,t') = <A(t)B(t')>,$ \qquad (9)

The Equation of the motion for this CF can be received from equations of the motion for operator, falling into (9):

$$i\frac{dA}{dt} = A\widetilde{H} - \widetilde{H}A$$

$$i\frac{dG}{dt} = <(A\widetilde{H} - \widetilde{H}A)B(t')>.$$

CF can usually be part of the right part of an equation of a higher order, than source CF. Equations of the motion for these more complex CF contain even more complex CF, and so on. The System of grabbing equations is understood. The Number of these equations is infinite. Unhooking is used for decision. Unhooking tears off the chain and brings it into a closed system of equations. The Equation is necessary to complete the initial conditions. We shall get this from commutative correlations. The Obvious characteristic of CF, resulting from a round-robin transposition of operators under the sign of borehole is taken into account:

<A(t)B(0)> = <B(-iβ)A(t)>, (10)

We shall consider the Hamiltonian of Fermi (Bose) gas

$$\widetilde{H} = \sum T_f a_f^+ a_f$$

where $f = (k, \sigma)$, σ – a spin index, k – impulse (for Bose gas $\sigma = 0$), $T_f = (k^2/2m) - \mu$, μ – a chemical potential, a^+_r, a_r – operators, satisfying to corresponding commutative correlations

$a^+_r, a_r + a_r, a^+_r = \delta\ a_r, a_r + a_r, a_r = 0$ (for Fermi)

$-a^+_r, a_r + a_r, a^+_r = \delta\ a_r, a_r + a_r, a_r = 0$ (for Bose)

We shall consider CF $G_f(t) = <a^+_f(t)a_{f(0)}>$. The equation of the motion for a^+_f is of the form of

$$i\frac{d}{dt}a_f^+ = -T_f a_f^+.$$

We shall get the equation for $G_f(t)$

$$i\frac{d}{dt}G_f = -T_f G_f.$$

The Decision of this equation is:

$$G_f(t) = G_f(0)\exp(iT_f t). \tag{11}$$

We take into account (10):

$G_f(-i\beta) = <a^+_f \beta(-i\beta) a_f(0)> = <a_f(0) a^+_f(0)>$.

We take into account (11):

$G_f(-i\beta) = G_f(0)\exp(\beta T_f)$.

$G_f(-i\beta) \pm G_f(0) = <a_f(0) a^+_f(0) \pm a^+_f(0) a_f(0)> = 1$ follows from commutative correlations (sign + pertains to Fermi, a – to Bose case. Now $G_f(0)\exp(\beta T_f) \pm 1) = 1$ and $n_f = <a^+_f(0) a_f(0)> = G_f(0) = [\exp(\beta T_f) \pm 1]^{-1}$.

We get the functions of the distribution for ideal Fermi- and Bose- systems. The Chemical potential μ we shall define from condition $\Sigma[\exp(\beta T_f)+1]^{-1} = N$.

Chapter 2

THE METHOD OF THE CONSTANT SELF-CONSISTENT FIELD USED FOR THE DECISION OF THE PROBLEM OF MANY SOLIDS

2.1. INTRODUCTION

We have realized, all physical values are expressed through CF (FG). A problem is reduced to calculation of average (B− an operator of the birth and destruction or a product of it)

$$<B(t)>= Sp\{B\rho(t)\}/Sp\{\rho(t), <B>_0 = \frac{Sp\{B\rho_0\}}{Sp\{\rho_0\}}$$

where

$$i\frac{d}{dt}\rho = [H,\rho], \quad H = H_0 - AF(t).$$

The Calculation of average is simplified for small indignation. Expression for matrix of density is found simply

$$\rho(t) = \rho_0 - i \int_{-\infty}^{t} \exp(-i(t-t')H^x)V^x(t')\rho_0 dt'$$

ρ_0 – the equilibrium distribution of Gibbs ($t = -\infty$).

$$[\rho_{0'} A] = i \int_0^{\beta} d\lambda \rho_0 \dot{A}(-i\lambda).$$

Then

$$\rho(t) = \rho_0 + \int_0^{\beta} d\lambda \rho_0 \int_{-\infty}^{t} \dot{A}(t'-t-i\lambda) F(t')dt'$$

The Periodic external power presents special interest
$F(t) = F_0 \cos\omega t$,
then
$$\rho(t) = \rho_0 + \int_0^{\beta} d\lambda \rho_0 \int_{-\infty}^{t} \dot{A}(t'-t-i\lambda) F(t')dt' = \rho_0 + \operatorname{Re} \int_0^{\beta} d\lambda \rho_0 \int_{-\infty}^{t} \dot{A}(-y-i\lambda) e^{-i\omega y} F_0 \exp(i\omega t)dt$$

Average importance is expressed:

$$<B(t)> = <B>_0 + \operatorname{Re} \int_0^{\beta} d\lambda \int_{-\infty}^{t} <\dot{A}(-y-i\lambda)B>] \exp(-i\omega y) dy F_0 \exp(i\omega t)$$

The Decision is reduced to calculation

$<A(-i\lambda)B(t_1)>$

condition $<A(-i\lambda)B(0)> = <B(-i\beta)A(-i\lambda)>$ is enough for determination of integration constants in equations for CF.

2.2. EQUATIONS FOR CF AND THEIR DECISIONS

The Fermi system with Hamiltonian shall consider

$$H = \sum T_f a_f^+ a_f - \frac{1}{2V}\sum I(ff') a_f^+ a_{-f}^+ a_{-f'} a_{f'}$$

$$f = \vec{k}, \sigma$$

$$T_f = \frac{k^2}{2m} - \mu$$

We write the equation for CF $G_f = <a^+_f(t) a_f(0)>$.

Then

$$i\frac{d}{dt}G_f = -T_f G_f + \frac{1}{V}\sum_{f'} I(ff') <a_{f'}^+(t) a_{-f'}^+(t) a_{-f}(t) a_f(0)>,$$

The Dynamic problem of calculating CF powerfully becomes complicated at the presence of the interaction. The Separate particles moving in these conditions are correlated. Power, caused by surrounding the particle, acts on the chosen particle. A Change in this complex field by some average field or molecular field is the simplest posterization. The Average field posterizations at condition of the coincidence of incorporated average field with average field, by means of its chosen particle acts on nearby (the self-consistent condition).

Such approach to the problem of many solids is very general. It is used in different systems. In our system:

$$<a_{f'}^+(t) a_{-f'}^+(t) a_{-f}(t) a_f(0)> = C_1(f') <a_{-f}(t) a_f(0)> - C_2(f)\delta_{f'-f} G_f + C_2 \delta_{ff'} G_f,$$

Constants C_1 and C_2 are possible to consider equal expressions $C_1 = <a^+_f a^+_{-f}>$, $C_2 = <a^+_{-f} a_{-f}>$. Constants C_1 and C_2 do not depend on time.
Sometimes C_1 is zero. Hartree-Fock approximation is used.

A pairing of operators is produced under one time. These pairing are considered true average. So $C_2(f) = n_f$. New CF appears when unhooking. CF is not a zero $\Gamma_f(t) = <a_f(t)a_f(0)>$. The New indication shall be entered for combinations with $C_1(f')$

$$L_f = -\frac{1}{V}\sum I(ff')C_1(f')$$

Combinations do not depend on time. For $\Gamma_f(t) = <a_f(t)a_f(0)>$ have an equation (V − strives to infinity. Members, proportional of delta function, disappear).

$$i\frac{d}{dt}\tilde{A}_f(t) = T_f\tilde{A}_f - L_f^*G_f$$

The Equation for $G_f = <a^+{}_f(t)a_f(0)>$ will take type

$$i\frac{d}{dt}G_f(t) = -T_f - L_f\tilde{A}_f(t).$$

We solve system:

$G_f(t) = A\exp(i\omega_f t) + B\exp(-i\omega_f t)$, where $\omega_f^2 = T_f^2 + [L_f]^2$. Magnitudes A and И are defined from initial conditions

$G_f(0) + G_f(-i\beta) = 1$, $\beta = 1/kT$, $\Gamma_f(0) + \Gamma_f(-i\beta) = 0$.

We solve the last system

$$G_f(0) = n_f = \frac{1}{2}(1 - \frac{T_f}{\omega_f}th\frac{\beta\omega_f}{2}) \quad \tilde{A}_f^* = -\frac{L_{f'}}{2\omega_{f'}}th\frac{\beta\omega_f}{2}$$

A condition of self-consistent

$$L_f = -\frac{1}{V}\sum l(ff')\tilde{A}_{f'}^*(0) = -\frac{1}{V}\sum l(ff')\frac{L_{f'}}{2\omega_{f'}}th\frac{\beta\omega_f}{2}$$

The integral equation allows decisions with $L_f = 0$ and $L_f \neq 0$. The Decision corresponds to superconducting condition when $L_f \neq 0$.

The Unhooking of specified type is systematically possible to generalize. A Derived on time is approximated by the expression in the manner of the final amount of some operators.

$$[A_i, H] = \Sigma k_{ij} A_j \tag{1}$$

Different ways of self-consistent conditions are used for determination κ_{nm}.

The remarkable feature of unhooking is the absence of the fixed vacuum, or the fixed basis. Switching properties of operators are used. Vectors of conditions are not entered into the theory. It follows that the vacuum becomes physical (symmetry more close to the decision).

Unhooking is exact for ideal systems, for example, for ideal gases. (1) is exact for these cases

For the Frelih Hamiltonian

$$H = \sum_{k\sigma}\frac{k^2}{2m}a_{k\sigma}^+ a_{k\sigma} + \sum \omega_q b_q^+ b_q + \sum A_q a_{k1\sigma}^+ a_{k2\sigma} r_q \; ,$$

$$r_q = b_q + b_q^+$$

A Partial diagonalization it is spent by this method (at $\kappa^2 = 0$). This method is useless at the width of a zone of conductivity distinct from zero. Care is always necessary. This approach leads to no physical collapse of free energy to infinite negative size for the some Hamiltonians. Consequentially, it would be the wrong use of CF in a steam room on small distances. CF reverts in negative infinity on small distances. Such non physical expression of CF breaks the thermodynamic stability of the system and leads to infringement of the second law of thermodynamics. This incorrectness in the expression of free energy has prominent features of false phase transition with root behavior in a thermal capacity, and can be wrongly interpreted as true phase transition. Contradictions with the second law of thermodynamics disappear, «phase transition» simultaneously disappearing as if to make CF small and positive. This can sometimes be a poor approach.

For example, exact additives to one and two-partial CF can be found by considering a problem about dispersion of easy particles on heavy particles.

$$\Delta f = \exp(-\varepsilon t), \Delta g = \theta(t_1)\theta(t - t_1)\theta(t)\exp(-\varepsilon(t - t_1)),$$

where ε a constant, $\theta(t)$ – the Hivisayd stage. We solve a problem by means of unhooking:

$$\Delta f_0 = \exp(-\varepsilon t), \Delta g_0 = \theta(t_1)\exp(-\varepsilon(t - t_1))$$

The member $\theta(t - t_1)\theta(t)$ is lost. The approach works poorly at $t_1 > t$ (greater distances between particles).

Approximations of a constant field can lead to incorrect results. They are unsuitable for systems with strong interaction.

The basic lack of a method of the self-coordinated field consists of replacement of a real fluctuating field with some averages and constants. Sometimes a constant field can be considered as an incidental size independent of time. Function of distribution set. The system of the equations is broken into an infinite set of equations with a various constant field. The average decision is made, averaged on distribution of molecular fields. The spectrum elementary возбуждений can be received from CF in the usual image. It will depend on a kind of function of distribution of incidental fields. Excitation will not be elementary. Averaging destroys poles. The spectrum will be not polar. Systems, which have no elementary stimulations (systems with strong interaction), are well described by this method.

2.3. The Conclusion

The method of a molecular field or approach of chaotic (incidental) phases, often yields incorrect results, as an infringement of laws of thermodynamics. It is necessary to supervise results and check their conformity to organic laws of physics. The method demands improvement.

Chapter 3

DYNAMIC AND STOCHASTIC SYSTEMS

3.1. INTRODUCTION

All interesting physics systems are complex and we should consider the use of CF. The infinite system of hooking equations of movement with CF higher order can be made for every CF. This infinite system completely defines required CF as an independent task of entry conditions. An Unhooking should take into account the contribution from all infinite chain in the maximal image. The basic contribution to averages brings only a part of times. We should uncouple CF correctly on this interval.

3.2. DYNAMIC AND STOCHASTIC SYSTEMS

All systems can be divided asymptotically into two classes: dynamic and stochastic. We shall show on an example of autocorrelation functions (with identical operators) G (t) = <A(t)A(0)>.

G(t) decreases after an exhibitor at greater times. On small times G(t) possesses conditions:

$$\dot{G}(0) = < \dot{A}(0)A(0) > \tag{1}$$

$$\ddot{G}(0) = -(< \dot{A}(0)\dot{A}(0) >) < 0 \tag{2}$$

The dynamic behavior of systems is carried out under influence (1), (2) on behavior of CF. If (1), (2) do not influence behavior of system G(t), stochastic systems arise. G(t) behaves as exp(-εt).

A satisfies the equation:

$$i\frac{dA(t)}{dt} = (-\omega_0 + M(t))A(t)$$

<M(t)> = 0

$$G(t) = G(0)\exp(i\omega t) < \exp\left\{i\int_0^t V(t_1)dt_1\right\} >.$$

$$\tau_c = \int_0^\infty \frac{<M(t)M(t+\tau)>}{<M(0)M(0)>} d\tau \qquad (3)$$

it is accepted for correlation time.

The case of prevalence of dynamic communication corresponds to a condition:

<M(0)M(0)>$^{1/2}$ τ_c >> 1 (4)

The case of prevalence of stochastic communication corresponds to a condition:

<M(0)M(0)>$^{1/2}$ τ_c >> 1 (5)

Averaging in G(t) can be lead for Gauss process:

$$G(t) = G(0)\exp(i\omega_0 t)\exp\left\{-\int_0^t (t-t_1) <M(0)M(t_1)> dt_1\right.$$

The case (4) corresponds to slowly varying function M(t).

$$G(t) = G(0)\exp(i\omega_0 t)\exp\left(-\int_0^t \frac{<M(0)M(0)>}{2}t^2\right),$$

$$\int_0^t (t-\tau)<M(0)M(\tau)>d\tau = t\int\int_0^\infty <M(0)M(\tau)>d\tau = t\tau_c <M(0)M(0)>$$

It corresponds to fast disappearance of dynamic communication. Dynamic communication prevails at t t<<τ_c. Stochastic communication prevails at t>>τ_c.

Transition of system to stochastic behavior does not occur all at once. The stochastic behavior is described by the Boltzmann equation and supposes the approach of a constant field. Thin stochastic layers of trajectories in phase space for small times are remote from each other by the invariant surfaces which are not allowing them to be agitated. Transition to merging stochastic layers arise with increase in time and there is a stochastic movement.

The phase space can be divided three areas. One area contains stochastic trajectories. The kinetic description and approach of a molecular field is fair for this area. This area is connected by trajectories of movement with the second area. The significant part of the second area is made with stochastic components of movement, but small islands of regular movement are available. Islands are connected by trajectories with stochastic areas. Kinetic equations and molecular field approach are not used in this area. The third area (dynamic) contains mainly regular movements. Regular field approach is unsuitable.

<M(0)M(0)> characterizes size of interactions in the system in relation to the chosen dynamic variable. If interaction is great, τ_c is great. Effects of dynamic communication are essential in this case. If A is the electronic operator, the M(t) is a phonon field (Freliha system). <M(0)M(0)>$^{1/2}$ $\tau_c = \gamma$. It varies from 3 up to 40. The system is dynamic (for organic semiconductors especially).

Unhookings with a constant field cannot lead to Gauss form G(t) at unhooking at any final stage. This form is presented often in the theory of a firm body.

The problem of strong and intermediate communication arises. Effects of dynamic communications play a defining role in dynamic systems for research of a power spectrum, thermodynamic properties, and the phenomena of carry. In the method of the self-coordinated field, the kinetic Boltzmann equation is insufficient. Semiconductors with small mobility (oxidation of many organic semiconductors and polymers) are an example of such systems. Spin diffusion, processes in shock waves, and free nuclear induction in crystals are difficult to study. We encounter strong displays of communication in these and other processes. $\tau_c = \omega^{-1}$ in systems with small mobility (where the model of jumps works well). ω – The Effective oscillatory frequency. The force acting on the carrier from an oscillatory subsystem of M(t) is proportional A. A – the effective size of interaction of carriers with a lattice.

$\gamma = <M(0)M(0)>^{1/2}$ $\tau_c = A/\omega$ can be great. The Gauss form of a signal (for a signal of a free induction) shows that the phenomenon occurs to an area of strong communication. How can these problems be solved? How can we use CF to define these problems? New approaches are necessary. The introduction of time-dependent self-coordinated fields is one such approach. This method is a generalization method of incidental phases (or a molecular field).

3.3. METHODS OF THE VARIABLE SELF-COORDINATED FIELD

The environment acts on the evolved particle only statically. It is the basic lack of methods of the self-coordinated field. At the first stage, it is necessary to consider effects of fluctuation and its consequence (infinite system of gearing equations). A Concept about a variable floor it is entered instead of a representation of the self-coordinated constant floor. The variable field can provide faster convergence in a number of consecutive approximations. Even zero approach will be good. Properties of ideal gas are lost for an unperturbed system. Another system (for example, polaron) gets out for unperturbed system. The elementary way of introduction of the self-coordinated field is direct unhooking. We shall consider an example, the Matsubara Hamiltonian

$$B = \sum_q A_q(b_q + b^+_{-q}); \quad C = \sum_q A_q(b_q + b^+_{-q}).$$

$$i\dot{a} = \omega_0 a + Ba, \quad i\dot{C} = \omega B + 2A^2 a^+ a,$$

$$iB = \omega C, \quad A^2 = \sum_q A_q^2$$

CF $<a^+(t)a(t_1)>$ we will research. We will write equations:

$$i\frac{d}{dt}<a^+(t)a(t_1)> = -\omega_0 <a^+(t)a(t_1)> - <B(t)a^+(t)a(t_1)>,$$

$$i\frac{d}{dt}<B(t)a^+(t)a(t_1)> = \omega_0 <Ba^+(t)a(t_1)> + <B(t)a^+(t)B(t_1)a(t_1)>.$$

Unhooking can be done. We tear off operators of phonons from electronic. The time scale is different for these operators.

$<B(t)a^+(t)B(t_1)a(t_1)> \approx <B(t)B(t_1)><a^+(t)a(t_1)>$

(1)

Expression $<B(t)B(0)>$ is easily calculated. It concerns to isolated phonons system. It is equal:

$$B(t)B(0) >= \left\{ \frac{4A^4}{\omega^2} + A^2 \left[Ne^{i\omega t} + (N+1)e^{-i\omega t} \right] \right.$$

$N = <b^+_q b_q> = (e^{\beta\omega} - 1)^{-1}$.

We search for the decision of a kind:

$<a^+(t)a(0)> = e^{-i\omega t}G(t)$.

We have for G(t):

$$\ddot{G}(t) + <B(t)B(0)> G(t) = 0 \qquad (2)$$

This is the Mathieu equation. It is not solved generally.
The exact decision is known and considered Hamiltonian. The decision is determined as direct diagonalization of Hamiltonian.

$$G_r(t) = \exp\left\{\frac{A^2}{\omega^2}\left[N(e^{i\omega t}-1)+(n+1)(e^{-i\omega t}-1)-i\omega t\right]\right\}$$

Exact $G_T(t)$ satisfies to the equation:

$$\ddot{G}_r + G\left[A^2\left[Ne^{i\omega t}+(N+1)e^{-i\omega t}\right]+\frac{A^4}{\omega^2}\left[N^2 e^{2i\omega t}+(N+1)e^{-2i\omega t}+1-\right.\right.$$

$$\left.\left.-2N(N+1)-2Ne^{i\omega t}+2(n+1)e^{-i\omega t}\right\}=0 \qquad (3)$$

The equations (2) and (3) coincide at small A/ω or small times. Size $4A^4/\omega^2$ is had for the last group of members at t = 0. The Unhooking (1) it is impossible to consider unhooking well. It is exact only on small times. The decision (2) looks like (at small times and heats (N>> 1)):

$$G(t) = \exp(-bt^{2/2})({}_1F_1(\frac{b-a}{4b},\frac{1}{2},bt^2)+i\frac{2A^2}{\omega}t\,{}_1F_1(\frac{3b-a}{4b},\frac{3}{2},bt^2)) \qquad (4)$$

${}_1F_1$ – degenerate hypergeometrical function, a = $2A^2N$, $b^2 = A^2\omega^2N$. We shall note square-law dependence on time (4). It cannot be received by a method of the constant self-coordinated field. We do unhooking:

<B(t)a⁺(t)a(0)>=<B><a⁺(t)a>,

We shall receive

$$G(t) \approx \exp(-i\frac{2A^2}{\omega}t),$$

We do unhooking

$<B(t)B(t)a^+(t)a(0)> \approx <B^2><a^-(t)a(0)>,$

$<C(t)a^+(t)a(0)> \approx <C><a^+(t)a(0)>,$

We shall receive

$$G(t) \approx \frac{1}{2}(1+\frac{A^2}{\omega\Delta})e^{i\Delta t} + \frac{1}{2}(1-\frac{A^2}{\omega\Delta})e^{-i\Delta t},$$

$$\Delta^2 = <B^2> = (A^2(2N+1) + \frac{4A^4}{\omega^2}$$

$\Delta^2 = <B> = (A^2(2N+1) + 4A^4/\omega^2$

Unhooking with a constant field cannot lead to dependence (4) at any final stage. It is possible to consider a field of variables with the subsequent self-coordinated definition. Then:

$<B(t)a^+(t)a(0)> = M(t)<a^+(t)a>,$

$<B(t)a^+(t)a(0)> = M_1(t)<a^+(t)a>.$

New fields (their definition) are entered. How to find them? We suggest considering approximately:

$<B(t)B(t)a^+(t)a(0)> = (M^2(t) + D)a^+(t)a>,$ (5)

$<C(t)B(t)a^+(t)a(0)> = (M_1M + E)<a^+(t)a>,$ (6)

D, E – constants which are from the requirement of satisfaction to boundary conditions for CF, the maximum order of which are approximated. Equations for M(t) and $M_1(t)$ (t) follow from the equations of movement for CF

$iM = \omega M_1 - D$, $iM_1 = \omega M + 2A^2 - E$.
$D = A^2(2N + 1)$, $E = A^2$, $M(0) = -2 A^2/\omega$, $M_1(0) = 0$.

We solve system and we shall receive

$M(t) = A^2/\omega(Ne^{i\omega t} - (N + 1)e^{-i\omega t} - 1)$

$$G(t) = \exp\left(\frac{A^2}{\omega^2}(N(e^{i\omega t} - 1) + (N+1)(e^{-i\omega t} - 1) - i\omega t)\right)$$

This exact is decision.

Chapter 4

THE METHOD OF APPROACH MD

4.1. A METHOD OF FUNCTIONAL DERIVATIVES

New methods of unhooking are necessary for improvement. The general way of reception through the functional derivative of classical fields is found by approximation. Fields are included in a Hamiltonian equation. We shall consider an example. The equations for CF follow from the equation of movement for operators:

$$i\frac{\partial}{\partial t}a(r,t) = -\frac{\nabla^2}{2m}a(r,t) + \int d^3r' V(r',t)a(r,t)$$

$$(i\frac{\partial}{\partial t_1} + \frac{\nabla_1^2}{2m})G_n(1,\ldots,n;1',\ldots,n') + \int d^3r_{n+1} V(r-r_{n+1})G_{n+1}(1,\ldots,n,r_{n+1}t_1;1',\ldots,n',r_{n+1}t_1) = 0$$
(1)

Classical fields $\xi(r,t)$ and $\eta(r,t)$ we shall enter. A Functional we shall define:

$$G(\xi,\eta) = 1 + \sum_1^\infty \frac{\xi(n)\ldots\xi(1)}{n!} G_n(1,\ldots,n',1',\ldots,n') \frac{\eta(1')\ldots\eta(n')}{n'!}$$

CF is possible to receive from this functional by means of differentiation.

$$G_n = \frac{\delta}{\delta\eta(n')} \cdots \frac{\delta}{\delta\eta(1')} \frac{\delta}{\delta\xi(1)} \cdots \frac{\delta}{\delta\xi(n)} G(\xi,\eta),$$

$$\xi = \eta = 0$$

The infinite system of the equations (1) can be written down in the form of:

$$((i\frac{\partial}{\partial t_1} + \frac{\nabla_1^2}{2m})\frac{\delta}{\delta\xi(1)} + \int V(1,2) \frac{\delta}{\delta\eta(2+0)} \frac{\delta}{\delta\xi(2)} \frac{\delta}{\delta\xi(1)} d^3r_2) G(\xi,\eta) = 0$$
(2)

This approach is easier, than (1). Its decision is received with use of Fourier transformations. Fourier transformations clean derivatives. Then we should integrate functionally.

The method of functional derivatives allows us to easily understand long unhookings. A method of a molecular field features decomposition of the functional in Taylor's number on classical fields which are limited to two members of decomposition. The degrees of phase transitions do not calculate any precise improvements of the molecular field. Essentially if a molecular field is made time-dependent, characteristics of phase transitions do not improve. Characteristics are somewhat better than in the approach of various variants of a molecular field.

Some decisions for extremals occur in a method of functional derivatives. At calculation of functionals the instruction is an opportunity for spontaneous infringement of symmetry. The Singular covariant decision is realized in case of phase transition. The invariant decision is received under the theory of indignations and Vick and Feynman techniques. The Covariant decision is received by techniques of quasi-average of Bogolyubov. The method of functional derivatives deals with quasi-average by cause presence of classical fields. It is a good approach for reception of covariant decisions.

Further, we use a method of functional derivatives at the formulation of the generalized method of the variable self-coordinated field.

4.2. A Method of the Variable Self-Coordinated Field.

Systems with strong interaction between particles are the basic object of physical theories. The most interesting phenomena occur here. It is necessary to conserve a part of the dynamic generalized variables and average likelihood characteristics appear. This type of dynamic generalized characteristics depends on roughness of averaging.

The free induction of spins was the first investigated system of such type. We shall consider the model of Frelich with a zero zone of conductivity and square-law interaction on phonons. The model solved a method:

$$H = Ea^+a + \omega\Sigma b_q^+ b_q + \Sigma A_q r_q a^+ a + \Sigma B_{qq1} r_q r_{q1} a^+ a \qquad (1)$$

$$r_q = b_q + b_q^+$$

The ways to break chains for CF stated earlier have no strict requirements. Internal criterions of quality of unhookings to receive it are not possible. The value of unhookings are discovered by comparing the received results with experimental data.

An effective way to break chains is found in this case of strong communication. The hypothesis on the basis of a principle of easing of correlation can be used. The hypothesis is reduced to the statement, that the infinite system of particles is well described by the final number of other particles. A final part is left. Other parts replace it with an effective field. The field depends on time. Difficulty consists in definition of this field.

The basic hypothesis consists of: the Subsystem gets out from $n + 2$ particles. They are in a floor of the other $(N - n - 2)$ particles. It is a field we shall designate through $M(x, t)$. The subsystem from $n + 1$ particles which are being floor $(N - n - 1)$ particles feels same field $M(x, t)$. This assumption is not exact. Particles $n + 2$ and $n + 1$ can use correlated motion. We consider the amendment on correlation. We achieve a new powerful way to break a chain of equations for correlation functions (CF) or Green functions (FG). The method has been applied to spin systems for the first time.

CF is necessary to calculate the definition of an electronic spectrum of system

$$G(t) = Sp(e^{\beta(F-H)} a^+(t) a(0)) = <a^+(t) a(0)>_0,$$

F – a free energy. We use functional derivatives to construct a closed loop of the equations for CF. Functional derivatives allow us to formally write exact equations acceptable for various approximations. We shall enter generalized CF:

G(t) = Sp(pa$^+$(t)a(0)) = <a$^+$(t)a(0)> = <p$'$a$^+$(t)a(0)>$_0$/<p$'$>

$$\rho' = T \exp(\int_0^\infty dt \sum u_q(t) r_q(t)),$$

$$\rho = e^{\beta(f-H)} \rho' / <\rho'>_0$$

We have conserved for generalized CF the same designation, as usual. It in it passes at $u_q(t) \to 0$. CF has physical sense at $u_q(t) \to 0$.

Introduction of classical fields $u_q(t)$ in the definition of CF allows us to formally close a chain of equations for CF by means of functional differentiation. We shall consider the equation:

$$(t) = iEG + i\sum A_q < Tr_q(t)a^+(t)a(0) > + \sum B_{qq_1} < Tr_q(t) r_{q_1}(t) a^+(t) a(0) > \quad (2)$$

We take the variational derivative from G(t), and we receive

$$G(t) = < Tr_q(t_1) a^+(t) a(0) > - < r_q(t_1) > G(t)$$

or

$$< Tr_q(t_1) a^+(t) a(0) > = (< r_q(t_1) > + \frac{\delta}{\delta u_q(t_1)}) G(t)$$

Similarly

$$< Tr_q(t_1) r_{q_2}(t_2) a^+(t) a(0) > = (< r_q(t_1) > + \frac{\delta}{\delta u_q(t_1)})(< r_{q_2}(t_2) > + \frac{\delta}{\delta u_{q_2}(t_2)}) G(t)$$

The equation (2) can be copied in the "closed" form

$$\frac{d}{dt}G(t) = iEG + i\sum A_q(<Tr_q(t)> + \frac{\delta}{\delta u_q(t)})G(t) + \sum B_{qq_1}(<r_{q_1}(t)> + \frac{\delta}{\delta u_{q_1}(t)})g(t)$$

(3)

We copy the equation in the integrated form. New functions $M_q(t)$ $D_{qq1}(t,t_1)$ are entered under formulas

$$(<r_q(t_1)> + \frac{\delta}{\delta u_q(t_1)})G(t) = G(t)(M_q(t_1;t) + \frac{\delta}{\delta u_q(t_1)})$$

(4)

Similarly

$$\prod_1^N (<r_q(t_1)> + \frac{\delta}{\delta u_i(t_1)})G(t) = G(t)\prod_1^N (M_q(t_1;t) + \frac{\delta}{\delta u_i(t_1)})1$$

$$D_{q_1 q_2}(t_1,t_2;t) = \frac{\delta M_{q_1}(t_1;t)}{\delta u_{q_2}(t_2)} = \frac{\delta M_{q_2}(t_2;t)}{\delta u_{q_1}(t_1)}$$

(5)

The equation (3) can be copied in the integrated form:

$$G(t) = G(0)\exp(iEt + \int_0 \sum A_q M_q(t_1;t) + i\sum B_{qq_1}(M_q(t_2;t) + D_{qq_2}(t_2,t_2;t))dt_2$$

(6)

Chapter 5

THE METHOD OF APPROACH MD (CONTINUATION)

5.1. A METHOD OF VARIABLE SELF-COORDINATED FIELDS (CONTINUATION)

Equations 6, 4 and 5 form the "closed" system of the functional equations. Functional $M_q(t_1,\{u_q\};t)$ is entered. The identical representation noted takes place earlier

$$<T\prod_i r_{q_1}(t_1)A(t)> = <TA>\prod_i (M_{q_1}(t_1;t) + \delta/\delta u_{q_1}(t_i))$$

(7)

Where T - the operator of chronological ordering, A - any combination of operators.

$$\frac{\delta}{\delta u_q(t)}) <TA\exp(\sum \int dt_1 u_q(t_1)r_q(t_1))>_0 / T\exp(\sum \int dt_1 u_q(t_1)r_q(t_1))>_0) =$$
$$= <TAr_q(t)\rho'>_0 / <\rho'>_0 - <TA>_0 <r_q(t)>_0 / <\rho'>_0^2$$

We migrate members and we write down

$$<TAr_q(t)> = (<r_q(t)> + \frac{\delta}{\delta u_q(t)})<TA> = <TA>(M_q(t;t) + \frac{\delta}{\delta u_q(t)})$$

If the operator ($<r_{q1}(t_1)> + \delta/\delta u_{q1}(t_1)$) acts in both parts of the expression, we receive ($<r_{q1}(t_1)> + \delta/\delta u_{q1}(t_1)$)$<TAr_q(t)> = <TAr_q(t)r_{q1}(t_1)>$

Thus

$$<Tr_q(t)r_{q_1}(t_1)a^+(t)a(0)> = (<r_q(t)> + \frac{\delta}{\delta u_q(t)})(<r_{q_1}(t_1)> + \frac{\delta}{\delta u_q(t)})(<r_{q_1}(t_1)> +$$

$$+ \frac{\delta}{\delta u_{q_1}(t_1)}G(t) = (<r_q(t)> + \frac{\delta}{\delta u_q(t)})G(t)(M_{q_1}(t_1;t) + \frac{\delta}{\delta u_{q_1}(t_1)} =$$

$$= G(t)(M_q(t;t) + \frac{\delta}{\delta u_q(t)})(M_{q_1}(t_1;t) + \frac{\delta}{\delta u_{q_1}(t_1)}.$$

(7) it is proved similarly.

Using the variational derivative from G (t) which is included in the infinite system of hooking equations for this model, it is possible to express all variational derivatives $M_q(t_1, \{u_q\}; t)$ in a compact way. The system cannot be solved precisely. It is possible to approximate.

Approximation is investigated. Only functional derivatives of the first order from $M_q(t_1, \{u_q\}; t)$ are kept. Other derivatives are equal to zero.

$$\frac{\delta^n}{\delta u_{q_1}(t_1)...\delta u_{q_n}(t_n)} M_q(t;t) = 0, \ n \geq 2$$

Unhooking is reduced to the following: two identical replacements are used. Functionals $M_q(t_1, \{u_q\}; t)$ и $D_{qq1} = \delta M_q/\delta u_q$ are entered. That is:

$<Tr_q(t_1)A> = <TA>M_q(t_1,t)$

$<Tr_q(t_1)r_{q1}(t_2)A> = <TA>(M_q(t_1;t)M_{q1}(t_2;t) + D_{qq1}(t_1,t_2;t))$

The Method of Approach MD (Continuation)

We leave only the first derivative from $M_q(t_1,\{u_q\};t)$ for CF with three or greater number of operators $r_q(t)$ from an exact parity (7) after action of the right part on (1). The second and third derivatives are rejected. It is unhooking. Unhooking reduces the infinite system of linear equations for CF to finalize the system of nonlinear equations after transition to a limit $u_q(t) \to 0$.

The equation for G (t) is written down already in the form of the formal decision through M_q and D_{qq1} (the formula (6)). Thus, the equation for $M_q(t;t)$:

$$(M_q(t;t) = <r_q(t)> + \frac{\delta}{\delta u_q(t)} \ln G(t) = \frac{\delta}{\delta u_q(t)} \ln G(0) + r_q(t) > +i \int_0^t \sum A_{q_1} D_{qq_1}$$

$$(t,t_1;t)dt_1 + (8) + i \int_0^t \sum B_{q_1 q_1}(M_{q_1}(t_2;t_2)D_{qq_2}(t,t_2;t) + M_{q_2}(t_2;t_2)D_{q_2 q_1}(t_1;t_2;t))dt_2$$

At the initial moment of time $G(0) = 1$, yields:

$$<r_q(t) - <r_q(t)>> = 0$$

Also

$<r_q(0)>_{uq \to 0} = -2Aq/\omega_q$

The equation for D_{qq1} should be received. If we take a variational derivative from (8), we receive:

$D_{qq1}(t_1,t_2;t) = D^0_{q1q2}(t_1 - t_2)\delta(q_1 + q_2) +$

$$+ i \int_0^t \sum B_{qq_2} D_{q_2 q}(t_2,t_3;t)D_{q_1 q}(t_1,t_3;t)dt_3 + i \int_0^t \sum B_{qq_2} D_{q_2 q}(t_2,t_3;t)D_{q_1 q}(t_1,t_3;t))dt_3$$

$$(9)$$

Causal FG for free phonons:

$D^0_{q1q2}(t_1 - t_2) = <Tr_{q1}(t1) r_{q2}(t2)> - <r_{q1}(t1)> <r_{q2}(t2)>$

FG matters at $u_q \to 0$:

$$D_{q1q2}(t) = (N_q e^{i\omega_q t} + (N_q + 1)e^{-i\omega_q t}),\ t>0.$$

The equation (9) will define size D_{q1q2} and the system becomes isolated at $u_q \rightarrow 0$. Do not enter other functions in the theory (except for G, M_q, D_{q1q2}). Equations are for G, M_q, D_{q1q2}.

5.2. AN EXAMPLE

Let's consider a system with Hamiltonian:

$$H = \sum_{1}^{1} E_n a_n a_n + H' + \sum_{1}^{2} k_n a_n^+ a_n r^2$$

H' – the Hamiltonian of solvent, r – the coordinate of reaction. We shall calculate CF
$G(t) = <a^+(t)a_2(t)a_2^+(0)a_1(0)>$
This CF describes the probability of transition 1→2 (from the first condition in the second). The equation looks like:

$$i\frac{d}{dt}G = EG + k <r^2(t)a_1^+(t)a_2(t)a_2^+(0)a_1(0)>.$$

$k = k_2 - k_1$, $E = E_2 - E_1$. Do not enter linear on coordinate terms in system, parameter $A = 0$. $M_q(t;t) = 0$, $D(t_1,t_2;t)$ as defined from (9). System is received:

$$G(t) = G(0)\exp(-iEt - ik\int_0^t D(t_1,t_1;t)dt),$$

$$M_q(t;t) = 0$$

$$D(t_1,t_2;t) = D^0(t_2-t_1) - 2ik\int_0^t D(t_2,t_3;t)D(t_1,t_3;t)dt_3$$

Let

$$D^0(t_1-t_2) = -id^2\exp(-\omega|t_2-t_1|),$$

The Spectrum of the environment (solvent) is relaxation. We search B
$$D(t_1,t_2;t) = -id^2\exp(-\omega t_2 + \omega t_1)\varphi(t),\ t_2 > t_1.$$
This is substituted in the equation for D. We receive:

$$\varphi(t) = 1 - 2kd^2\int_0^t e^{-2\omega t_2}\varphi^2(t_2)dt_2$$

We differentiate on t the received expression. We receive the equation:

$$\varphi(t) = 2\omega - 2\omega\varphi(t) - 2kd^2\varphi^2(t),$$

The equation is easily solved and yields:

$$\varphi(t) = \frac{1}{2kd^2}(-\omega + \frac{d}{dt}(In(e^{\Omega t} + ce^{-\Omega t})))$$

We substitute it in expression for G. We define G in the form of:

$$G(t) = G(0)\exp(-iEt)\exp(-kd^2\int_0^t \varphi(t_1)dt_1) = G(0)(\frac{|1+c|}{|e^{\Omega t}+e^{-\Omega t}|e^{-\omega t}})^{\frac{1}{2}}\exp(-iEt)$$

It is difficult to receive this result in the usual way of summation of diagrams (the way of turning of numbers is not clear). This expression for G is exact and describes a relaxation of electrons in the weakening environment. G(t) decays exponentially at t $\to\infty$. The Behavior of G(t) is more complex in intermediate times.

Chapter 6

THE GENERALIZED MODEL OF MATSUBARA

We shall consider application of an offered method of unhooking in concrete models. We choose a generalized model of Matsubara with Hamiltonian.

$$H = \Sigma E_n a^+_n a_n + \Sigma \omega_q b^+_q b_q + \Sigma A_{qn} a^+_n a_n r_q / r_q = b_q + b^+_{-q}$$

This model finds sufficient application and describes radiation, not adiabatic transitions; carry of electrons and protons, carry of heavy ions, constants of speed of reactions and others. CF is necessary to calculate the definition of an electronic spectrum of system.

$$G_{mn}(t) = Sp(e^{\beta(F-H)} a^+_m(t) a_n(0)) = <a^+_m(t) a_n(0)>_0$$

We shall consider generalized CF for construction of the closed equations by means of variational derivatives:

$$G_{mn}(t) = Sp(p a^+_m(t) a_n(0)) = <a^+_m(t) a_n(0)>,$$

where

$$p = e^{\beta(F-H)} p^/ / <p^/>_0,$$

$$\rho' = T \exp \int_0^\infty dt \sum u_q(t) r_q(t)$$

We have now:

$$\frac{d}{dt}G_{mn} = iE_m$$

$$\frac{d}{dt}G_{mn} = iE_m G_m + i\sum A_{qm} <r_q(t)a_m^+(t)a_n(0)>,$$

$$\frac{\delta G_{mn}}{\delta u_q(t_1)} = <r_q(t_1)a_n^+(t)a_m(0)> - <r_q(t_1)>G_{nm}$$

$$<r_q(t_1)a_n^+(t)a_m(0)> = (\frac{\delta}{\delta u_q(t_1)} + <r_q(t_1)>)G_{nm}$$

We substitute the last expression in the equation of movement. We receive the closed equation:

$$\frac{d}{dt}G_{nm} = iE_n G_{nm} + i\sum A_{qn}(\frac{\delta}{\delta u_q(t)} <r_q(t)>)G_{nm},$$

We shall enter new functional M_q under the formula:

$$(\frac{\delta}{\delta u_q(t_1)} + <r_q(t_1)>)G_{nm} = G_{nm}(M_q(t_1;t,nm) + \frac{\delta}{\delta u_q(t_1)})1 = G_{nm}(M_q(t_1;t,nm)$$

Then

$$\frac{d}{dt}G_{mn} = iE_n G_{nm} + i\sum A_{qn}M_q(t;t,nm)G_{nm}(t)$$

From here

$$G_{mn}(t) = G_{nm}(0)\exp(iE_n t + i\sum \int_0^t A_{qn} M_q(t_1; t_1, nm) dt_1$$

On the other hand

$$M_q(t_1;t,nm) = <r_q(t_1)> + \frac{\delta}{\delta u_q(t_1)} \ln G_{nm}(t)$$

$$= <r_q(t_1)> + \frac{\delta}{\delta u_q(t_1)}(\ln G_{nm}(0) + iE_n t + i\sum \int_0^t A_{q_1 n} M_{q_1}(t_1;t_1,nm) dt_1)$$

$G_{nm}(0)$ from u_q does not depend, E_n too. We have the equation for definition M_q (a chain of the equations).

$$M_q(t_1;t,nm) = <r_q(t_1)> + i\frac{\delta}{\delta u_q(t_1)} \sum \int_0^t A_{q_1 n} M_{q_1}(t_3;t_3,nm) dt_3$$

We break this chain. We assume that:

$$\frac{\delta^2 M_{q_2}(t_3;t,mn)}{\delta u_q(t_1) \delta u_{q_1}(t_2)} = 0 \tag{2}$$

Higher variational derivatives are also equal to zero. This is our major assumption. We receive:

$$D_{q_1 q}(t_1;t_3,nm) = \frac{\delta M_{q_2}(t_1;t,mn)}{\delta u_{q_1}(t_2)} = <Tr_q(t_1) r_{q_1}(t_2)> - <r_q(t_1)><r_{q_1}(t_2)>$$

$$< iE \int_0^t A_{q_2 m} \frac{\delta^2 M_{q_2}(t_3;t_3,mn)}{\delta u_q(t_1) \delta u_{q_1}(t_2)} dt_3$$

We use a hypothesis (2)

$$D_{q_1q}(t_1,t_2;t_3,nm) = <Tr_q(t_1)r_{q_1}(t_2)> - <r_q(t_1)><r_{q_1}(t_2)>$$

The last expression is Green's causal function of free fields of phonons. Its value results are above:

Size

$$M_q(t_1;t,nm) = <r_q(t_1)> + i\sum \int_0^t A_{q_1n} D_{q_1q}(t_1,t_3;t_3,nm)dt_3$$

$<r_q(t)> = -2Aq/\omega_q$

Sizes of M and B are new and basic objects of the theory. All characteristics of system are expressed from them. We shall find out their physical sense. M_q describes some effective field acting on evolved particles. It depends on a kind of CF system, utilizing the size of interaction between particles, temperatures, time (oscillatory t_1 and electronic t).

Size D is a generalized causal FG of phonon. It depends on not only two times of an oscillatory subsystem (t_1 and t_2), but also from time t, describing a condition of an electronic subsystem. It depends on type of system, kind CF, parameters of Hamiltonian and temperatures.

The physical sense of prospective unhooking considers that there is very little indirect interaction between phonons through electrons. Obvious kinds of D can be received in case of linear interaction in the form of:

$D_{q1q2}(t_1,t_2;t, mn) = D^0_{q1q2}(t_1 - t_2)$.

D is equal to other expressions in other cases. Size D^0 is known and has already been established.

$D^0_{q1q2}(t) = (N_q e^{i\omega qt} + (N_q + 1)e^{-i\omega qt})\delta(q - q_1)$, t>0.

We substitute it in (3) and (4). We receive:

$$D_{q_1q_2}(t_1,t_2;t,mn) = D_{q_1q_2}(t_1 - t_2)$$

Expression coincides with exact.

The Generalized Model of Matsubara

We shall show that by means of M and D it is possible to precisely describe systems with Hamiltonian of generalized Matsubara. We shall calculate all the maximum variational derivatives from M and we shall show that they are equal to zero at $u_q \to 0$. We receive:

$$D^0_{q_1 q_2}(t) = (N_q e^{i\omega qt} + (N_q + 1)e^{-i\omega qt})\delta(q - q_1), \ t>0.$$

We should be convinced that the integrated member from the right is equal to zero. We shall consider the expression:

$$M_q(t_1; t, nm) = -(A_{-qn}/\omega_q)(-1 + N_q e^{i\omega qt} - (N_q + 1)e^{-i\omega qt}),$$

$$G_{nm}(t) = G_{nm}(0)\delta_{nm} \exp(iE_n t - i\sum \frac{|A_{qn}|^2}{\omega_q}t +$$

$$\sum \frac{|A_{qn}|^2}{\omega_q^2}(N_q(e^{i\omega qt} - 1) + (N_q + 1)(e^{-i\omega qt} - 1))),$$

We decompose to an exhibitor in a number, we consider, that a^+_n enter in $\hat{A}$, $(a^+_p a_p)^n$. Expression corresponds in an equivalent manner:

$$\frac{\delta^2 M_{q_1}(t_1; t, nm)}{\delta u_{q_2}(t_2)\delta u_{q_3}(t_3)} = <T(r_{q_1}(t_1) - <r_{q_1}(t_1)>)(r_{q_2}(t_2) - <r_{q_2}(t_2)>)$$

$$(r_{q_3}(t_3) - <r_{q_3}(t_3)>) + i\int_0^t \sum A_{q_4 n} \frac{\delta^3 M_{q_4}(t_4; t_4, nm)}{\delta u_{q_1}(t_1)\delta u_{q_2}(t_2)\delta u_{q_3}(t_3)}dt_4$$

New operators $\beta q = b_q + a_q$, enter. a_q - Constant number. We receive:
Be sure that the outer integral one on the right is zero. For this purpose, let us consider the first expression:

$$Sp(e^{-\beta H}\hat{A}) = Sp(\exp(-\beta \sum(E_n + \sum A_{qn} r_n + \sum \omega_q b_q^+ b_q) a_n^+ a_n)\hat{A})$$

Z – the constant. We can do the same operation for time dependence of operators $\hat{A}$. Time evolution and averaging of operators occurs on Hamiltonian of unbound particles.

Vick's theorem is carried out for such cases. Factoring the exponent in the series, and taking into account that a_n^+ are in $\hat{A}$ and that $(a_p^+ a_p) = a_p^+ a_p$, this expression can be rewritten in equivalent way.

$$Sp(e^{-\beta H}\hat{A}) = Sp(\exp(-\beta \sum (E_n + \sum A_{qn} r_n + \sum \omega_q b_q^+ b_q) a_n^+ a_n))\hat{A})$$

enters in each member. $<\Delta>$ is equal to zero, all expression to equal zero means. It is carried out for all odd averages; hence all of them are equal to zero. All even variational derivatives from M_q will have an integrated member. The integrated member for a functional derivative of the third order is equal:

I = $<T\Delta(1)\Delta(2)\Delta(3)\Delta(4)>$ -2$<T\Delta(1)\Delta(2)><T\Delta(3)\Delta(4)>-2<T\Delta(1)\Delta(3)><T\Delta(2)\Delta(4)>-2<T\Delta(1)\Delta(4)><T\Delta(2)\Delta(3)>$

Vick's theorem is applied to the first member, we receive 1 = 0. The integrated member and all odd derivatives on classical fields equal zero. . We continue the same calculations until we receive the equation:

$$Sp(e^{-\beta H}\hat{A}) = ZSp(\exp(-\beta \sum \omega_q \beta_q^+ \beta_q)\hat{A})$$

We consider that the system contains N particles. We receive:

$$\frac{\delta^{N-1} M_{q_1}}{\delta u_{q_2}(t_2)...\delta u_{q_{N-1}}(t_{N-1})} = i\int_0^t \sum A_{q_N n} \frac{\delta^N M_{q_n}(t_N ...)}{\delta u_{q_1}(t_1)...\delta u_{q_N}(t_N)} dt_N$$

Only the member is distinct from zero. It proves the accuracy of unhooking MD through our Hamiltonian.

Chapter 7

RESEARCH OF NONEQUILIBRIUM DYNAMIC PROBLEMS

We shall estimate the accuracy of a method of unhookings and we shall consider CF for calculation of chemical reaction, hopping conductivity and properties of polaron small radius in this lecture.

How to define the accuracy of results? To estimate accuracy of a deviation, we should take the exact result and compare it to the received result. Which exact result should we take? One normal way of achieving exact results is known as the theory of indignations. It is necessary for us to consider all orders on a parameter of indignation.

We shall consider a binodal Hamiltonian (the index 1 at Hamiltonian and CF means, that the system with linear interaction is considered).

$$H = \Sigma E_n a^+_n a_n + \Sigma \omega_q b^+_n b_q + \Sigma A_{qn} a^+_n a_n r_q,$$

$$r_q = b_q + b^+_{-q}, n = 1,2$$

We shall calculate CF through which kinetic coefficients are expressed (for a change):

$$G_1(t) = <a^+_1(t)a_2;(t)a^+_2(0)a_1(0)>$$

We shall lead calculation of this CF all over again under the theory of indignations, then by method MD. We shall write $G_1(t)$ to represent the interaction. We shall accept:

$H_1 = H_0 + V$,
where

$H_0 = \Sigma E_n a^+{}_n a_n + \Sigma \omega_q b^+{}_n b_q$, $n = 1, 2$.

– not indignant system (ideal gas). Indignation $V = \Sigma A^{(n)}{}_q r_q a^+{}_n a_n$, $n = 1, 2$. $G_1(t)$ is possible to write in the form of (using approach $(a^+a)^n = a^+a$):
$G_1(t) = <Ta^+{}_1(t)a_2(t); a^+{}_2(0)a_1(0)T\exp(-i\int \Sigma A_q r_q(t_1)dt_1>$,
Where $A_q = A_q^{(1)} - A_q^{(2)}$, and dependence of operators from time is defined by the operator of Hamiltonian H_0:

$a^+{}_1(t)a_2(t) = \exp(i(E_1 - E_2)t) a^+{}_1(0) a_2(0)$.

We apply representation to the received expression:

$$<T\prod_{i-1}^{N} r_{q_1}(t_i)\hat{A}(t) >=<T\hat{A}>\prod_{i-1}^{N}(M_{q_i}(t_1;t...) + \frac{\delta}{\delta u_{q_i}(t_i)})1$$

Then:

$$G_1(t) = G_0(t)\exp(-i\int_0^t \sum A_q(M_{0q}(t_1;t) + \frac{\delta}{\delta u_q(t_1)})dt_1)1,$$

where

$$M_{0q}(t_1;t) = <r_q(t_1)> + \frac{\delta}{\delta u_q(t_1)} \ln G_0(t),$$

$$G_0(t) = G_0(0)\exp(i(E_1 - E_2)t)$$

We shall take advantage of disentangling exhibitors in $G_1(t)$ of that:

Research of Nonequilibrium Dynamic Problems

$$M_{0q}(t_1;t) = <r_q(t_1)>_1 \quad <r_q(t)>_{u_{q=0}} = -\frac{2A^{(1)}_q}{\omega_q}$$

$$<r_Q(t_1)> = <r_Q(t_1)>_{u_{Q=0}} + \sum \int_0^{t_1} D^0_{qq_1}(t_1-t_2) u_{q_1}(t_2) dt_2$$

M_{0q} is linear on u_q, к $G_1(t)$. Identity Veyl is possible to apply. For our case

$$\exp(A+B) = \exp A \exp B \exp(-\frac{1}{2}[\hat{A},\hat{B}])$$

$$\hat{A} = -i\int_0^t \sum A_q M_{0_Q}(t_1;t) dt_1 \quad \hat{B} = -i\int_0^t \sum A_q \frac{\delta}{\delta u_q(t_1)} dt_1$$

We represent an exhibitor in $G_1(t)$ through identity Veyl and we operate with the operator B on unit. We receive:

$$[\hat{A},\hat{B}] = \int_0^t\int_0^t dt_1 dt_2 \sum A_q A_{q_1} D^0_{qq_1}(t_1-t_2)$$

We apply MD approach:

$$G_1(t_1) = G_0(t)\exp(-i\sum A_q(M_{0_q}(t_1;t) - \frac{1}{2}\int_0^t\int_0^t \sum A_q A_{q_1} D^0_{qq_1}(t_2-t_1) dt_1 dt_2)1$$

where

$$M_{iq}(t_1;t) = <r_q(t_1)> + \frac{\delta}{\delta v_q(t_1)} In G_1(t) = r_q(t_1) > -i\int_0^t \sum A_{q_1} \frac{\delta}{\delta v_q(t_1)} M_{1_{q_1}}(t_2;t_2) dt_2$$

— linear function u_q. The Hypothesis MD is carried out exactly. We received earlier the resulted exact expression for $G_1(t)$.

We shall consider more complex Hamiltonian

$H_2 = H_1 + V$, $V = \Sigma B_{qq1}^{(n)} r_q r_{q1} a_n^+ a_n$, H_1 is certain above. We shall consider

$G_2(t) = <a^+_1(t)a_2(t); a_2^+(0)a_1(0)>$.

$$H_2 = H_1 + V, V = \sum B_{qq1}^{(n)} r_q r_{q1} a_n + a_n + a_n$$

$$G_2(t) = <a +_1(t); a_2 + (0)a_1(0)>$$

The index «2» means that the system with square-law interaction is considered. The index «1» means the system with linear interaction, «0» — the system without interaction.

We continue representation of interaction. We receive:

$G_2(t) = <a^+_1(t)a_2(t); a_2^+(0)a_1(0) \exp(-i\int r_q(t_1) B_{qq1} r_{q1}(t_1) dt_1)>$, $B_{qq1} = B^{(2)}_{qq1} - B^{(1)}_{qq1}$.

Hamiltonian H_1 defines development on time and a matrix of density. We bear operators from under a sign on an average. We receive:

$$G_2(t) = G_1(t)\exp(-i\int_0^t \sum (M_{1q}(t_1;t) + \frac{\delta}{\delta v_q(t_1)}) B_{qq_1}(M_{1q}(t_1;t) + \frac{\delta}{\delta v_q(t_q)}) dt_1)1$$

It is necessary for us to show, that $\ln G_2(t)$ is square-law functional from $u_q(t)$. We shall enter vectors for convenience of calculations:

$M = \{M_{1q}(t_1;t)\}$, $\Delta = \{\delta/\delta u_q(t_1)$

and a square-law matrix B with elements $B_{q1q2} = B_{q1q2}\delta(t_1 - t_2)$.

$G_2(t)$ will become: $G_2(t) = G_1(t)\exp(-i(M + \Delta)B(M + \Delta))$.

We shall calculate switching properties of new operators:

MB = BM, ΔB = BΔ, ΔM = MΔ + D_1, $[M^2, M\Delta] = -2D_1M$, $[M\Delta, \Delta^2] = -2\Delta D_1\Delta$, $[M^2, \Delta^2] = -2D^2_1 - 4MD_1\Delta$,
Matrix D_1 does not depend from $u_q(t)$ and has elements:

$$D_{1q_1q_2}(t_1, t_2; t) = \frac{\delta M_{1q}(t_1; t)}{\delta v_{q_2}(t_2)}$$

Calculation of switchboards show that they all generate finite-dimensional algebra with elements M^2, $M\Delta$, Δ^2, 1. Other operators do not appear. It is possible to search for decomposition in the form of:
$\exp(\lambda(M + \Delta)B(M + \Delta)) = \exp(A_0 + MA_2M)\exp(MA_1\Delta)\exp(\Delta A_3 \Delta)$,

where A_i - infinite-dimensional matrixes. We differentiate this expression on λ and we use conditions of switching. We equate factors at identical operators. We receive system of the equations for definition A_i (i = 0,1,2):

$B = A'_2 - 2 A'_1 A_2 D_1 + 4 A'_3 e^{-2A_1 D_1}(A_2 D_1)^2$,
$B = A'_3$,
$2B = A'_1 - 4A'_1 e^{-2A_1 D_1} A_2 D_1$,
$BD_1 = A'_0 - 2A'_1 e^{-2A_1 D_1} A_2 D_1^2$.

We have for exact expression G_2:

$$M_{2q}(t_1; t) = <r_q(t_1)> + \frac{\delta}{\delta U_q} \ln G_1(t) +$$

$$+2\int_0^t dt_3 dt_2 M_{1q1}(t_2; t) A_{2q_1q_2}(t_2; t_3; t) D_{1q_2q},$$

$$D_{2q_2q_1}(t_2; t_1; t) = \frac{\delta <r_{q_1}(t_1)>}{\delta U_{q_2}(t_2)} + \frac{\delta^2}{\delta U_{q_2}(t_2)\delta U_{q_1}(t_1)} \ln G_1(t) +$$

$$+\int_0^t dt_3 dt_4 \sum D_{1q_3q_4}(t_2; t_3; t) A_{2q_3q_4}(t_3; t_4; t) D_{1q_4q_1}(t_4; t_1; t)$$

The right part of the last expression does not depend on u_q and the condition is precisely satisfied

$$\frac{\delta D}{\delta \upsilon_q(t)} = 0$$

This is the proof of accuracy of unhooking for this purpose of Hamiltonian. The MD unhooking approach is not exact since cubic anharmonicity, as the algebra of operators M^3, $M^2\Delta$, $M\Delta_2$, Δ^3, includes all degrees M^n, Δ^n. The specified breakage is in this case the confidant. Nevertheless, this approach is very good.

This approach has been applied to spin systems, to liquid, and to polymers and has yielded good results.

Chapter 8

CONDUCTIVITY IN POLYMERS

8.1. INTRODUCTION

Jump conductivity on some orders of film is observed with a number of polymers having threadlike structure, at weak change of external conditions (pressure, an electric field). The law of Ohm is carried out outside of area of jump. Interest in mechanisms of switching in polymers is caused by the polymeric substances which are the most productive for use in microelectronics where thin spending film with anisotropic and nonlinear characteristics is required.

An attempt to find superconductivity in polymeric systems is most interesting. This interest amplifies the opening of superconductivity in ceramics. Properties of ceramics and polymers are similar. The mechanism of superconductivity in ceramics is unknown. One superconductor candidate is polyacetylene. Synthesis of polyacetylene is well mastered, and physical properties are widely investigated. Polyurethanes, oxidized polypropylenes, and polyphenilethylenephtalides, are substances with similar properties.

Polyphenilethylenephtalides possess especially interesting properties. They are insulators or semiconductors in the basic condition. They become high-conductivity connections at switching. The volt-ampere characteristic of a spending condition appears bistable and looks like the butterfly or a hysteresis. We shall begin studying polyarylenphtalydes.

8.2. CONDUCTIVITY IN POLYARYLENPHTALYDES

Conductivity in polymers is carried out on channels that are located by chaotic image and penetrate a film of polymer.

The mechanism of fast formation and the disappearance of spending channels represent big interest. Models of type of metal moustaches are offered. They do not explain all properties of such channels. We shall consider the mechanism of formation of such a channel, connected by transformation of a polymeric circuit. We explain the majority of experimental results known at the present time: reversible jump conductivity on some orders at the appendix small uniaxial pressure approximately in 10^2 Pascal and (or) an electric field; tolerance of a film to hydrostatic pressure; absence of effect in thick tapes and its occurrence with the addition of pounded glass.

8.3. A STRUCTURE OF A POLYMERIC CIRCUIT

The reason for switching a polymeric film from not spending condition in spending, under action of small indignation, is covered in a structure of a monomeasured part which for the given class of polymers (polyarylenphtalydes) is in two updates: 1) the atom of carbon has sp^3 hybridization (the plane of benzene ring of lateral group is approximately perpendicular the basic circuit, the circuit of interface of monomeres along the basic circuit is broken off). 2) The Atom of carbon has sp^2 hybridization and provides interface between the monomeasured parts of a circuit.

Gradual transformation of not spending condition in spending is shown on Figure 8.1.

The mechanism:

Polymer consists of parts 5 and 6.

External influence breaks off one communication in initial parts of polymer. All polymers consist of parts 5 and 6. Two initial parts look like 3 and 4.

Two initial parts will be transformed in spending, becoming, specified on 1 and 2. Initial parts pass on a circuit to places 3 and 4, and 5 and 6 continue a circuit.

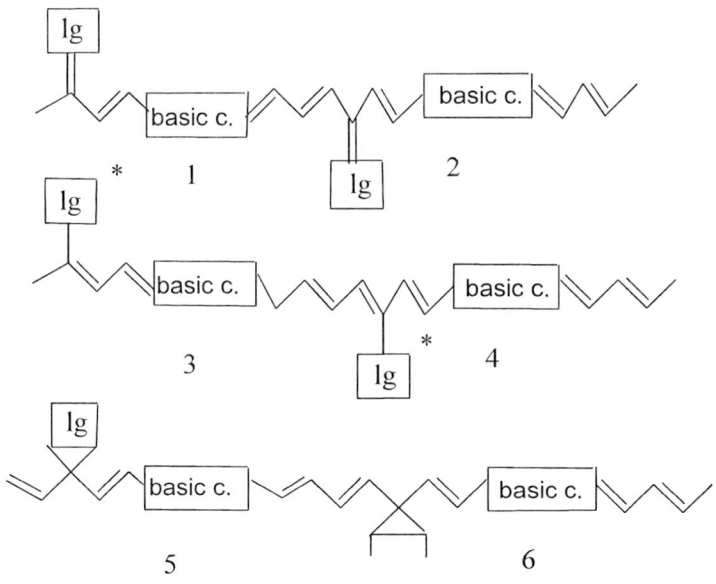

Figure 8.1. The scheme of distribution of soliton on polymer.

The Number of parts of type 1 and 2 accrues at the left and spending "tail" is formed. The transitive area of polymer from two parts of type 3 and 4 follows spending "tail". Not spending area from parts of type 5 and 6 follows further. This area gradually decreases in sizes.

The atom of carbon has sp^3 hybridization. The plane of benzene ring of the lateral group (in our case a phtalyd radical) is perpendicular the basic circuit ($\varphi = 0$). φ − A corner of turn of lateral group. The atom of carbon in the spending bridge has sp^2 hybridization. The plane of lateral group (lg) becomes a parallel basic circuit: $\varphi = \pi/2$. The structure of bridges is characterized by a corner φ, which has not only the specified extreme values, but also all intermediate.

Our system is described by Hamiltonian:

$$H = j/2 \sum_{n,g} \left[\sin^2(\varphi_{n+g} - \varphi_n) \right] +$$

$$+ \sum_{n,g} \left[\sin^2 \left(\frac{\varphi_{n+g} + \varphi_n}{} \right) \right] (a^+_{n+g} a_n + a^+_n a_{n+g})$$

$$+ \sum_n \left[\frac{M_n^2}{21} + \frac{u}{2} \sin^2 2\varphi_n + A \sin^2 \varphi_n \right]$$

φ_n – a corner of turn of unit n, a^+_n, a_n – operators of a birth and destruction of electron on unit n. Other sizes are numerical parameters.

Turn of a plane of lateral group in a spending condition adds in a π–environment one electron. Population of an environment increases for size $sin^2\varphi_n$.

The equations of movement follow from this Hamiltonian:

$$I \frac{d}{dt} \varphi_n = [\varphi_n H] = \frac{1}{I} \frac{\partial}{\partial \varphi}$$

We investigate changes of a unit of structure, and band structures of polymer, by means of these equations.

8.4. A SOLITON, REFORMATIVE THE POLYMER FROM NOT SPENDING CONDITION IN SPENDING

We have in semi classical approach, turns on a corner φ_n classical, after averaging on quantum variables:

$$\frac{\partial}{\partial t} \frac{d}{\partial o_n} = \frac{J_1}{1} \sin 2\phi_n \sum_g (a^+_{n+g} a_{n+g} a_n + a^+_n a_{n+g})$$

$$\sum \sin 2(\varphi_n - \varphi_{n+g})(a^+_{n+2g} a_n + a^+_n a_{n+2g})$$

$$\frac{H}{i} \sin \varphi + \frac{A}{1} \sin 2\varphi_b$$

Conductivity in Polymers 51

$$\sum (H_{20} + a^+_{n-2g}) \sin^2(\varphi_{n+g} - \varphi_n)$$

We consider, that <...> poorly depends on time and coordinates and we enter designations

$\alpha = \Sigma_g <(a^+_n a_{n+g} + h.c.)>$,

$\beta = <(a^+_n a_{n+2g} + h.c.)>$.

Prospective weak dependence is possible in continuous approach. We shall make it, believing that, $\varphi_n(t) + \varphi_{n+g}(t) = 2\varphi(n,t)$, a $\varphi_n(t) - \varphi_{n+g}(t)$ is small, which we shall leave only in the first order.

If $n = x/a$, we have:

$$\frac{\partial^2 \varphi(x,t)}{\partial t^2} = -\frac{u}{I} \sin 4\varphi(x,t) - \frac{A}{I} \sin 2\varphi(x,t) -$$

$$-\frac{J_1}{I} \alpha \sin 2\varphi(x,t) + \frac{4J_2 \beta a^2}{I} \frac{\partial^2}{\partial x^2} \varphi(x,t)$$

We shall enter scales of length and time so that factors at the second derivative on coordinate and at $sin4\varphi(x,t)$ are equal to unit. We have then:

$$t_0 \sqrt{\frac{1}{4u}}, \quad a = \sqrt{\frac{u}{j_2 \beta}}$$

Expression (3) will become:

$$\frac{\partial^2 y}{\partial t} - \frac{\partial^2 y}{dx^2} + \sin y = -\lambda \sin \frac{y}{2}$$

$y = 4\varphi$, $\lambda = (A + J_i)/u$ и далее везде $t = t/t_0$, $x = na$.

This equation is the equation double Gordon-sinus. When u it is great (usually) it it is possible to approximate by using the standard equation of Gordon-sinus in zero approach.

We have for electronic operators similarly, believing $sin\varphi_{n+g} = sin\varphi_n$

$$i\frac{d}{dt}a_n^+(t) = -2j_1t_0 \sin^2 \varphi(x,t)(a_{n+1}^+(t) + a_{n-}^+ 1(t))$$

Chapter 9

CONDUCTIVITY IN POLYMERS (CONTINUATION)

9.1. INTRODUCTION

Various types of equations are realized depending on boundary and entry conditions

$$\frac{\partial^2 y}{\partial t^2} - \frac{\partial^2 y}{\partial x^2} + \sin y = -\lambda \sin \frac{y}{2} \qquad (1)$$

$$\frac{\partial^2 y}{\partial t^2} - \frac{\partial^2 y}{dx^2} + \sin y = 0 \qquad (2)$$

Decisions can be oscillating nearby $\varphi = 0$ and $\varphi = \pi/2$. They do not lead to reorganization зонной structures of polymer as they keep the average constant value of a corner and describe oscillatory movements of the lateral group in a spending or isolating condition of a monomeasured part.

The most important decision (2) is:

$$\varphi(x,t) = arctg(\exp(\pm 2y\sqrt{\frac{4u - a + \beta}{a + \beta}}(x - x_0 - v_s t)))$$

$\gamma = 1/$, v – a constant, v_s – speed of soliton. From the decision (3) follows, that at infinite negative coordinate and a sign "minus" at γ a corner φ is equal $\pi/2$ and with increase x up to positive infinity monotonously varies up to zero. Another decision, at a sign «+» at, describes change of a corner from zero at infinite negative coordinate up to $\pi/2$ at x equal to positive infinity. The choice of decisions is defined by boundary conditions on the ends of a polymeric molecule.

9.2. Boundary Conditions

We shall consider external influence on a part of the polymeric molecule adjoining the surface of an electrode where φ is equal $\pi/2$ for the task of boundary conditions. The received values extend on a circuit according to expression (3). Charges Z of lateral groups of a polymeric circuit cooperate with an electrode with potential energy $Ze/2x_i$. x_i – distance from a charge up to an electrode when the lateral group has a minimum of potential energy. About these minima (their two: the point x and x_1) the lateral group makes movements with frequency, but does not pass from one minimum in another. Energy of activation Ea needs to be overcome for transition. Energy of activation varies in external electric floor E and at the appendix of uniaxial P. The Activation energy can be presented in the form of:

$$E^{\pm}_a = E_a - a_1 P \pm a_2 E$$

$a_1 = Z^2 e^2 \delta / 2x$ – the factor of pressure, δ – the factor of linear compression, $a_2 = \sqrt{2E_a/\omega}$ – factor of a field. The number of spending channels in steps varies in process of achievement a condition:

$$E^i_a - a_1 P \pm a_2 E \leq kT,$$

E^i_a –a set of activation energies, k – the Boltzmann's constant. We shall consider the influence of external influences on the formation of boundary conditions. Frontier groups have a corner $\varphi = 0$ (Figure 9.1, a). The appendix of pressure or a field to an electrode translates the frontier lateral group of an equilibrium condition with a minimum of energy u in a point x in a new equilibrium condition with potential energy u_1 in a point x_1, thus there is a distortion of a potential curve in such a manner that u_l becomes less u (Figure 9.1, b).

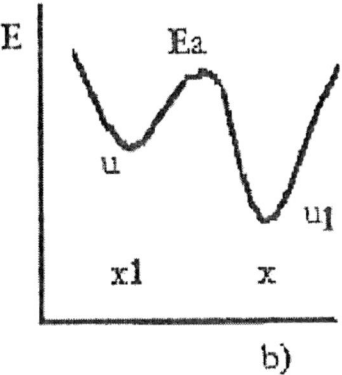

Figure 9.2. The scheme of potential energies in a vicinity of an electrode.

Communication C–O is broken off. The frontier lateral group borrows the position characterized by a corner $\varphi = \pi/2$ (Figure 9.1, a). It defines a boundary condition for the equation (1), generating the soliton. The potential curve comes back to the initial form (Figure 9.1, a) at removal of pressure (or fields) when the value of energy u in a point x is less than the value u_l in a point x_l and the corner of frontier lateral group accepts value $\varphi = 0$, generating the anti- soliton (Figure 9.1, a, b).

9.3. THE THEORY OF INDIGNATIONS FOR THE EQUATION (1).

Decisions of the indignant equation Gordon-sinus are sufficiently studied, in detail. Such decisions are searched in the form of soliton not indignant equation with modulated $(x + t)/2 = \xi$, $(x - t)/2 = \eta$, in which the equation becomes:

$$y_{\xi\eta} = \sin y + \lambda \sin (y/2)$$

We search for the decision in the form of $y = 4\operatorname{arctg} f$, $f = \exp\theta$.
We have the decision for not indignant equation

$$\theta^{(0)} = \pm\gamma[(1-v)\xi + (1+v)\eta].$$

This decision is deformed at the presence of indignation, that is speed of soliton and the form of an impulse varies. These changes are described by the equation

$$(\theta_{\xi\eta} - \lambda/2)\operatorname{cth}\theta = \theta_\xi \theta_\eta - 1.$$

We search for the decision in the form of

$$\theta = \theta(w), \quad w = a\xi - \gamma x_0 + \eta/a$$

We receive the equation

$$\theta_{ww} - \theta^2_w \, th\theta = \lambda/2 - th\theta \quad \text{или} \quad w = \int d\theta (1 + \lambda/2 \, sh2\theta)^{-0.5}$$

We receive in the first order on λ

$$w = \int d\Theta \left(1 + \frac{\lambda}{2} sh2\Theta\right)^{-\frac{1}{2}}$$

Higher orders are calculated under the formula of fusion

$$w = \int d\Theta \left(cch^2\Theta - sh^2\Theta + \frac{\lambda}{2} sh2\Theta\right)^{-\frac{1}{2}}$$

We analyze last expressions. The revolting member changes the form and speed of soliton, but does not destroy it.

For reception of the decision in this expression, we shall lower the last member, which plays a role only in transitive area. This area has the small sizes. We approximate the second member step function. We shall enter slow time $\tau = vt$. We shall consider time as our parameter.

$$i \frac{d}{td} G_{nm} = \varepsilon_n G_{nm} + \frac{J_1}{2}(G_{n+1,m} + G_{n-1,m})$$

$$n \leq \tau / b,$$

Conductivity in Polymers (Continuation)

$$i\frac{d}{dt}G_{nm} = \varepsilon_n G_{nm} = \varepsilon G_{nm}$$

Each unit of a polymeric chain is isolated from the neighbor on a circuit at $n > \tau/b$. Repeatedly, a degenerate band is formed in this part of the polymer with energy

$$E_n = \varepsilon, n = N + 1, \ldots\ldots$$

All $N = \tau/b$ units form the interfaced chain at $n \leq \tau/b$. We decompose in this interval correlation function (CF) in number of Fourier. We consider that CF is equal to zero. We have on borders of an interval:

$$i\frac{d}{dt}G_{km} = (\varepsilon + j_1 \cos(\frac{2\pi k}{N}))G_{km} \quad k=0,\ldots,N-1$$

$$G_{km} = G_{km(0)} \exp\left\{-it\left[\varepsilon + j_1 \cos(\frac{2\pi k}{n})\right]\right\}$$

The parameter exhibitors do not depend on an index m, power levels for this part of a polymeric circuit is described by expression

$$E_k = -J_1 \cos(\frac{2\pi k}{N}) \quad k=0,\ldots N-1$$

Half of conditions of this zone are borrowed electrons, which earlier were in an σ–environment. Occurrence of spending chains cause broadening strips of absorption π–π*. transition to size J_1 and occurrence of low-frequency absorption due to intraband transitions.

The crack for intraband transitions depends on slow time and the process of increase in a chain of interface as movement of kink decreases, aspiring in a limit to zero. It is described by expression:

$$\Delta(t) = J_1 2\pi b/vt$$

This expression allows the defining speed of soliton. The gauge of variable pressure with frequency Ω place on border of contact of polymer with an electrode for definition v. Increase of pressure generates soliton, and through a half-cycle, anti-soliton is generated at the reduction of pressure, extending with the same speed and limiting the interfaced part of polymer in size $2\pi v/\Omega$. The crack depends on frequency and is equal $\Delta(\Omega) = J_1 b\Omega/v$. $v = J_1 b\Omega/\Delta$. All sizes in this expression are known or are easily defined.

9.5. The Mechanism of Switching

We receive the basic result and establish the fact of formation of the running lonely wave leading transformation of a polymeric molecule from not spending condition in spending and back. The given class of polymers were not obvious in advance. It has allowed us to offer the model of formation of spending channels. The mechanism follows:

The Bridge atom of carbon in a polymeric circuit in the basic, not spending condition has sp^3 – hybridization. The plane of the lateral ring attached to given atom, is perpendicular the basic circuit of polymer ($\varphi = 0$). Reorganization of an electronic condition begins with turn of a lateral ring on a corner $\pi/2$, break of communication C – O in a vicinity of an electrode under influence of external influences of formation of double communications C = C instead of unary in the basic circuit. Reorganization of communications C – C in C = C also occur inside of unit and obviously are not reflected in model. The result, visible and considered, in the model is the break of communication C – O in the next part with formation biradical excited state.

Change of a corner φ from zero up to $\pi/2$ without attenuation extends on a circuit of polymer through an electronic subsystem, owing to the indirect interactions defined in parameters α and β. It occurs as follows. Bond is broken off. The molecule borrows energetically favorable position with a plane of a lateral ring under a corner $\pi/2$ to a plane, a perpendicular main circuit. It leads to occurrence in a π –environment one new electron which up to это was in an σ – environment, to its interaction with electron on bridge atom of carbon with its transformation from the form>C= in the form −C−. Turn of double bond on a corner $\pi/2$ occurs without change of hybridization of bridge atom. Replacement of double bonds C=C on unary C−C and on the contrary occurs in the basic circuit of polymer because of such turn. This process grasps two additional monomeasured parts. The biradical condition moves on one step, the interfaced circuit increases due to reduction of not interfaced part.

These of instability are observed in polymers at interaction of an electric field and uniaxial pressure. The received results allow us to explain experiments for polyarylenphtalydes, with the effect of switching with jump of specific resistance from 10^{14} up to 10^3 ohm−cm. It is observed in electric fields in thin tapes (thinner than 3 micrometers). The same effect is observed at the appendix of uniaxial pressure. Thus, external influence transforming properties of polymers is not enough. For example: threshold uniaxial pressure of transition is less 10^2 Pascal. Experiments show that sometimes the spending condition of polymer "is remembered". It can be connected with fixing turn on impurity or others heterogeneities. Deenergizing (occurrence of the second decision), in this case, can be carried out by stronger external influence: short electric impulses of a current, external influence, shift of an electrode.

Generation of solitons - process improbable. The transformed polymeric circuits will be little, but their conductivity is great and defines conductivity of a tape. The observable superficial density of spending channels (number of channels on unit of the area of an electrode) depends on conditions of preparation of the sample and character of external influences and reflects probability of generation soliton in the given conditions in experiments on switching.

9.6. OPTICAL PROPERTIES

The new strip arises at transition of polymer in spending condition in optical spectra with a maximum in the field of J_1. The strip sharply breaks aside greater energies and has a long tail aside of small energies. $J_1 \sim 1$ eV, occurrence of a new strip leads to coloring of polymer. Strips of absorption which were observed at not spending polymer, at transition in spending condition are spread and get satellites from high and low energies, distant from a maximum on sizes $J_1/2 + J_0$ and $J_1/2 + J_0$, where J_0 −width of the bottom zone. From here, there is a transition at absorption. Usually, it is $J_0 > J_1/2$ for the majority of organic molecules.

Concentration of spending molecules is small, but the absorption factor of them is strongly increased because of the big length of interface. It promotes the display of a spectrum of absorption of the interfaced molecules. These changes in spectra were observed experimentally. Two new strips of absorption were observed at 1.05 eV and 3.2 eV. Polymer in not spending condition had maxima of absorption at 4 eV and 4.5 eV, and in a spending condition, they have borrowed accordingly positions 4.1 eV and 4.9 eV.

9.7. ELECTRIC PROPERTIES

The given class of polymers has the volt-ampere characteristic (figure 9.2.) . The tape is in not spending condition and submits to the law of the Ohm on a site 0−1. Intensity of a field in a point 1 reaches such size, that the lateral group appears capable to break a barrier. The tape passes in spending condition on the way 1−2. The spending condition and applicability of the law of the Ohm are kept up to a point 3 at reduction of intensity of a field by virtue of that energy of activation of transition $x_1 - x$ is less, than energy of activation of transition $x - >x_1$. Then transition in not spending condition follows. The law of the Ohm is again carried out up to a point 5 if after achievement of a point 2 to continue to raise intensity of a field. Lateral groups come into effect in a point 5, where there are additional spending channels and a current grows up to value in a point 6. Further, all processes repeat. Switchings occur similarly in the sample prepared for a spending condition, but to deenergize spending channels and transition of a tape in not spending condition.

The described processes do not depend on polarity that is reflected in the left part of figure 9.2. The direction of a current varies at change of a sign on intensity of a field. The form of dependence is kept.

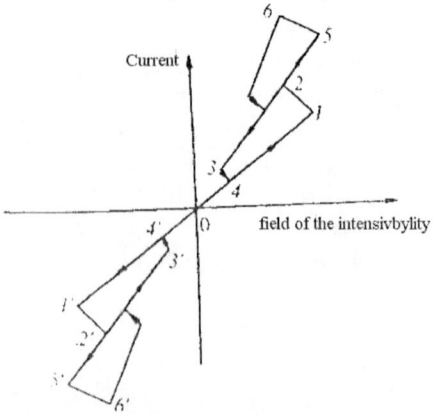

Figuire 9.2. Volt-amperic characteristic.

Chapter 10

A SPECTRUM OF ELEMENTARY ELECTRONIC CONDITIONS IN POLYARYLENPHTALYDES WITH TWO SOLITON EXCITATIONS

10.1. INTRODUCTION

Polyarylenphtalydes (PAph) can be a good model for studying a role of solitons. It is a polymer that has a quasi-crystalline structure, consisting of crystals and amorphous parts. We shall consider a circuit that passes both these phase parts. We shall be interested in the change of electronic conditions eventually due to moving two-solitons. These changes in behavior of a spectrum are shown in electro conductivity and can be experimentally measured. The Band structure can be investigated, applying optical methods.

Research of the influence of solitons on an electronic spectrum of polymeric systems of type the PAph is of interest. It allows us to find out the mechanism of transition metal− dielectric in such systems. Electronic levels start to depend on time, which has been last from the beginning of switching during this phase transition. This dependence can be used for research of development of process.

10.2. SOLITON EXCITATION OF POLYMER

The Modeling Hamiltonian, describing a chain of polymer, looks like [1]

$$H = j_1/2 \sum_{n,g} (\sin^2 \phi_n + \sin^2 \phi_{n+g})(a^+_{n+g}a_n + a^+_n a_{n+g})$$

$$+ J_2 \sum_{n,g} \sin^2 (\varphi_{n+g} - \varphi_n)(a^+_{n+2g} a^+_n a_{n+2g})$$

$$+ \sum_n \left[-\frac{1}{21} \frac{\partial^2}{\partial \phi_n^2} + \frac{u}{4}(1 - \cos 4\varphi_n) + \frac{A}{2}(1 - \cos 2\varphi_n) \right]$$

Where φ_n – a corner of turn of unit n, a^+_n, a_n – operators of a birth and destruction of electron on unit n, other sizes are numerical parameters.
Now we have

$$i\frac{d}{dt}\varphi_n = [\varphi_n, H] = \frac{1}{/}\frac{\partial}{\partial \varphi_n},$$

$$i\frac{d}{dt}(\frac{1}{/}\frac{\partial}{\partial \varphi_n}) = \frac{J_1}{/}\sin 2\varphi_n \Sigma_g (a^+_{n+g}a_n + a^+_n a_{n+g}) +$$

$$+ \frac{2J_2}{/} \Sigma \sin 2(\varphi_n - \varphi_{n+g})(a^+_{n+2g}a_n + a^+_n a_{n+2g}) +$$

$$+ \frac{u}{/}\sin 4\varphi_n + \frac{A}{/}\sin 2\varphi_n$$

$$i\frac{d}{dt}a^+_n = [a^+_n, H] = -J_1 \Sigma (\sin^2 \varphi_n + \sin^2 \varphi_{n+g})a^+_{n+g} + ...$$

We have a quasi-classical approach averaging on quantum variables, including turns on a corner φ_n classical, and electronic movements quantum:

$$\frac{\partial^2 \varphi_0(t)}{\partial t^2} = [[\varphi_n, H], H] = -\frac{u}{l}\sin 4\varphi_n - \frac{A}{l}\sin 2\varphi_n -$$

$$-\frac{J_1}{l}\Sigma_9 <(a_n^+ a_{n-g} + h.c.)> \sin 2\varphi_n -$$

$$-\frac{2J_2}{l}\Sigma g <(a_n^+ a_{n+2g} + h.c.)> \sin 4\varphi_n - \frac{A}{l}\sin 2(\varphi_n - \varphi_{n+g})$$

We enter for brevity designations:

$a = \Sigma g <(a_n^+ a_{n+q} + h.c.)>$, $\beta = <(a_n^+ a_{n+2q} + h.c.)>$, $n = x/a$

We are limited continuous approximation, copy (2) in the form of:

$$\frac{\partial^2 \varphi(x,t)}{\partial t^2} = -\frac{u}{l}\sin 4\varphi(x,t) - \frac{A}{l}\sin 2\varphi(x,t) -$$

$$-\frac{J_1}{l} a \sin 2\varphi(x,t) + \frac{4J_2 \beta a^2}{l}\frac{\partial^2}{\partial x^2}\varphi(x,t) \quad (3)$$

We shall enter scales of length and time so that factors at the second derivative on coordinate and at $sin4\varphi(x/t)$ are equal to unit. We have:

$$t_0 = \sqrt{\frac{1}{4u}}, \quad a = \sqrt{\frac{u}{J_2 \beta}}$$

Expression (3) becomes:

$$\frac{\partial^2 y}{\partial t^2} - \frac{\partial^2 y}{\partial x^2} + \sin y = \lambda \sin\frac{y}{2}, \quad (4)$$

where $y = 4\varphi$, $\lambda = -(A + J_1)/u$ and further everywhere $t = t/t_0$, $x = na$.

We have for electronic operators, believing that $sin\varphi_{n+q} = sin\varphi_n$

$$i\frac{d}{dt}a_n^+(t) = -2J_1t_0 \sin^2 \varphi(x,t)(a_{n+1}^+(t) + a_{n-1}^+(t)) \tag{5}$$

We shall enter new function $F = tg(y/4)$. $Siny/2 = 2F/(1 + F^2)$. We calculate $siny$, receive the equation for F. For $\lambda = 0$. It becomes simpler and becomes

$$\left(1+F^2\right)\left(\frac{\partial^2 F}{\partial t^2} - \frac{\partial^2 F}{\partial x^2} + F\right) - 2F\left(\left(\frac{\partial F}{\partial t}\right)^2 - (\frac{\partial F}{\partial x})^2 + F^2\right) = 0 \tag{6}$$

This equation has the private decision of a kind:

$$tq\frac{y}{4} = \frac{sh\frac{O_1 - O_2}{2}}{a_{12}^{1/2} ch\frac{O_1 + O_2}{2}} \quad a_{12} \neq 0, \tag{7}$$

$$O_i = \frac{a_i}{|a_i|}\frac{1}{\sqrt{1-v_i^2}}(X - X_i - v_i t),$$

$$V_i = \frac{1-a_i^2}{1+a_i^2}, a_{12} = \frac{(a_1 - a_2)^2}{(a_1 + a_2)^2}.$$

Four independent numerical parameters x_1, x_2, a_1, a_2 vary in a complex plane from $-\infty$ to $+\infty$ on a real and imaginary coordinate. 0π–Kinkit is possible to receive at a special choice x_1, x_2, a_1, a_2 from (7). We believe $a_1 = a_2 = a(cosm + Isinm) = f_{F+1} a, x_1 = x_2 \cdot{}^* = x_R + I x_1, a^2 = a^2{}_R + a^2{}_I, x^2 = x_R^2 + x_I^2$. We receive

$$tg\frac{y}{4} = ctgm\frac{sinO_1}{chO_R} \tag{8}$$

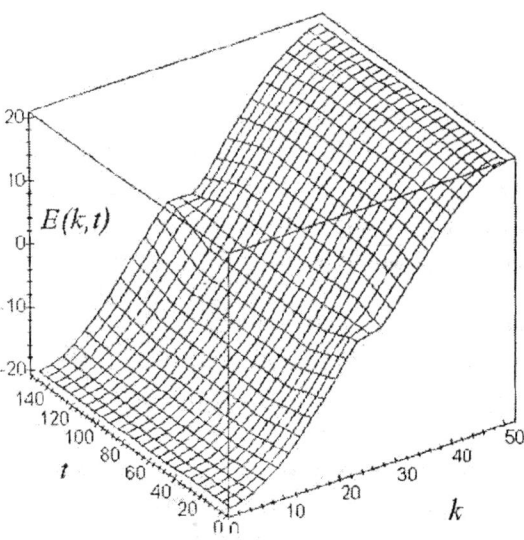

Figure 10.1. Change of band structures in due course. Parameters are accepted $2J_1t_0 = 10$, $t_1 = 60$, $v_1 = 0.19$, $v_2 = 0.21$, $x_1 = 29$, $x_2 = -34$. Conditions UZ disappear, pass in conditions ShZ, then on the contrary.

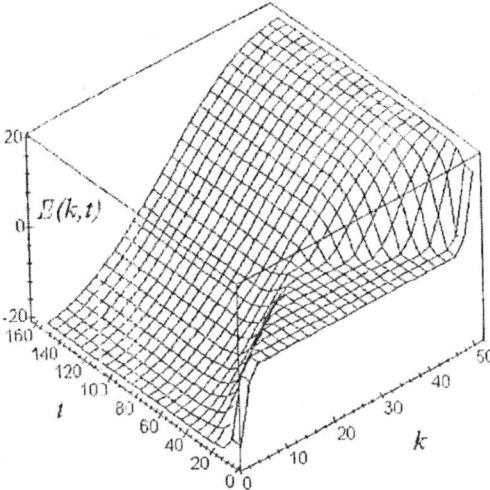

Figure 10.2. Change of band structures in due course. Parameters are accepted $2J_1t_0 = 10$, $t_1 = 10$, $v_1 = 0.60$, $v_2 = -0.46$, $x_1 = 29$, $x_2 = -23$. Conditions UZ disappear under the linear law from time.

$$o_r = \frac{\cos m}{\sqrt{1-v^2}}(x-vt) - \frac{1}{2}a_r x_r$$

$$o_1 = \frac{\cos m}{\sqrt{1-v^2}}(t-xv) - \frac{1}{2}a_1 x_1, \quad v = \frac{1-a^2}{1+a^2}$$

Decisions with divided variables represent the greatest interest. They can be received from (7), but to deduce from (6) is easier. We search for decisions in the form of $tg\varphi = f(x)/g(t)$. We shall receive the equations for f and g:

$$\left(\frac{df}{\partial t}\right)^2 = \mu f^4 + (1+\lambda)f^2 - \nu$$

$$\left(\frac{dq}{\partial t}\right)^2 = -\mu g^4 + \lambda g^2 + \nu$$

where, μ, λ, ν – any constant numbers. We have two sets of decisions at $\mu = 0$:

$$f = \pm ch\gamma(x-x_1), \quad \pm sh\gamma(x-x_1)$$

$$g = \pm \frac{1}{v} sh\gamma v(t-t_1), \quad \pm \frac{1}{v} ch\gamma v(t-t_1)$$

$$\lambda = \frac{1}{\sqrt{1-v^2}}$$

v – speed of soliton. We have expression for bisoliton decisions:

$$tg\varphi_1(x,t) = \pm v \frac{ch\gamma(x-x_1)}{sh\gamma v(t-t_1)}$$

Expression has no interpretation. Expression

$$tg\varphi_2(x,t) = \pm v \frac{sh\gamma(x-x_1)}{ch\gamma v(t-t_1)}$$

describes soliton–anti-soliton and anti-soliton–anti-soliton collisions. The top sign corresponds to soliton–soliton collision, and the bottom sign corresponds to anti-soliton–anti-soliton.

We can search for decisions in the form of:

$$tg\varphi = \frac{q_1(t)}{f_1(x)}$$

We receive two more sets of decisions:

$$g_1 = \pm\frac{1}{v}ch\gamma v(t-t_1) \;,\; \pm\frac{1}{v}sh\gamma v(t-t_1)$$

$$f_1 = \pm sh\gamma(x-x_1) \;,\; \pm ch\gamma(x-x_1)$$

These decisions lead to bisoliton decisions:

$$tg\varphi_3(x,t) = \pm\frac{1 ch\gamma v(t-t_1)}{v sh\gamma(x-x_1)}$$

This decision has no evident interpretation

$$tg\varphi_4(x,t) = \pm\frac{1 sh\gamma v(t-t_1)}{v ch\gamma(x-x_1)}$$

and corresponds to soliton–anti-soliton dispersion. Occurrence of these excitements in a system of linear polymers changes an electronic spectrum.

10.3. CHANGE OF BAND STRUCTURES SOLITONS

This question cannot be analytically investi indignation. Numerical methods are used. We shall c by the formula (7). For it:

Sin
$$^2\varphi = -\frac{sh^2\dfrac{0_1-0_2}{2}}{a_{12}ch^2\dfrac{0_1+0_2}{2} + sh^2\dfrac{0_1-0_2}{2}}$$

We substitute this expression in (5). Dependence of this size on time (but not from coordinate) is weak. We do inverse Fourier transform on fast electronic time:

$$E(t)a_m^+(t,E) = -2J_1t_0\sin^2\varphi(x,t)\{a_{n+1}^+(t,E) + a_{n-1}^+(t,E)\} \quad (10)$$

Values are defined from the decision of the equation rather than E:

$$\det\left|E\delta_{nn_1} + 2J_1t_0\sin^2\varphi(na,t)(\delta_{n+1n_1}+\delta_{n-1n_1})\right| = 0$$

Size E depends on time, speeds v_1 and v_2 of solitons, a relative positioning of solitons, defined in the parameters x_1 and x_2..

We shall consider a case when speeds v1 and $_{v2}$ are directed to the different party and have various values. The behavior of levels of energy is presented in figure 10.1. Strong dependence of energy on number of a condition is available for a part of conditions (figure 10.1.). These conditions form a wide zone (IP) on which conductivity is carried out. This zone does not cover everything, and is only a part of units of a polymeric chain.

ShZ is available at $|t-t_1|$. UZ appears at $|t-t_1|>0$.. The number of conditions in UZ grows with growth $|t-t_1|>0$. (figure 10.3).
The behavior of power levels from time essentially varies for bisolitons φ_3 φ_4. We have for φ_3:

$$\text{Sin}^2 \varphi_3 = \frac{ch^2 yV(t-t_1)}{ch^2 yV(t-t_1) + V^2 sh^2 yx}$$

The eventual increase in number of conditions UZ occurs in an electronic zone. The maximum is reached at $|t-t_1| = 0$. Reduction of their number occurs then. The received picture reminds us of "pie" (figure 10.4). Such fluctuation in time for electronic conditions of polymer can be critical for many processes. ShZ is at $|t-t_1| = 0$, but has a small number of conditions.

We have for φ_4:

$$\text{Sin}^2 \varphi_3 = \frac{sh^2 yV(t-t_1)}{v^2 ch^2 yX + sh^2 yv(t-t_1)}$$

Change of a zone is same, as in the previous case. However, ShZ is absent at $|t-t_1| = 0$ (figure 10.5).

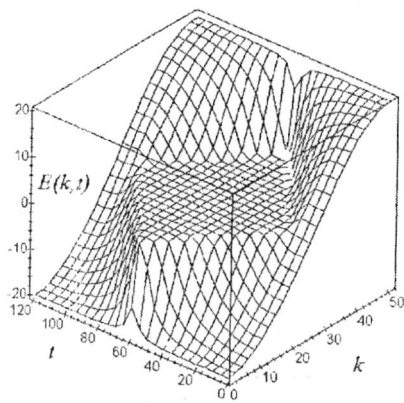

Figure 10.5. Change of band structures in due course. Parameters are accepted $2J_1 t_0 = 10$, $t_1 = 60$, $v_1 = -v_2 = v = 0.50$, $x_1 = x_2 = 26$. There is full disappearance of condition ShZ.

Bisolitons φ$_3$ φ$_4$ are arranged so that polymer is in a spending homogeneous condition at $|t-t_1| = \infty$. This condition is defining for them and takes place for the basic time interval. Not spending condition with UZ exists limited time, small in comparison in due course existence ShZ of a condition. The process occurs so that the number of conditions in a spending zone suddenly decreases and there are conditions with zero energy (UZ), and after short time conductivity again is restored. All these processes depend on parameters solitons, but restriction on time of existence of a mix of conditions UZ and ShZ always takes place.

$$\sin^2 \varphi_5 = \frac{\sin^2 \omega(t-t_1)}{f^2 ch^2 yx + \sin^2 \omega(t-t_1)}$$

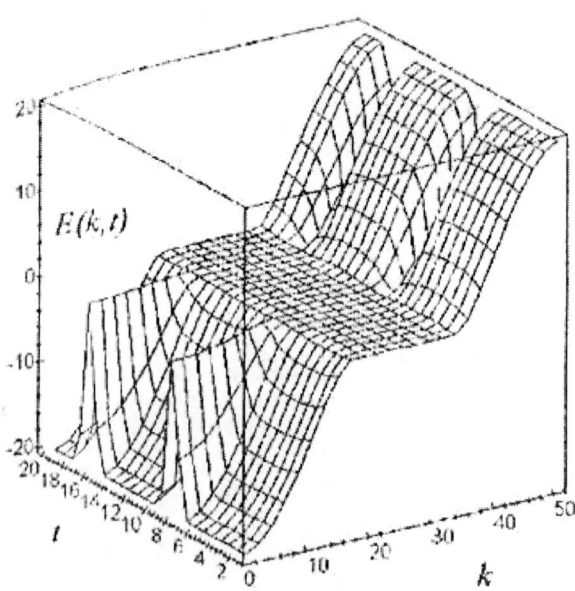

Figure 10.6. Change of band structures in due course for breather. Parameters are accepted $2J_1t_0 = 10$, $t_1 = 16$, $v = 0.50$, $\omega = \pi/8$, $f^2 = 0.5$. The number of conditions UZ does not vary, conditions ShZ completely disappear in special points on time.

The Band structure of polymer for breather depends periodically on time not varying on a number of levels in UZ. Their number is defined by size f^2. ShZ and periodically disappears at performance of a condition $\omega |t-t_1| = n\pi$,